Shargiil Bashir is a seasoned banker and experienced leader with over 20 years of experience. He has led various aspects of Corporate Governance, Strategy, Risk Management, Risk Assurance, and Sustainability throughout multiple countries.

He is a native Dane and started his career with Danske Bank in Copenhagen. Here, he held various senior leadership positions based primarily in Copenhagen and London.

In 2020, he joined First Abu Dhabi Bank where he. is the Chief Sustainability Officer and Executive Vice President. In this role, Shargiil is responsible for developing, leading, and implementing the bank's ESG strategy and initiatives. He was among the main drivers behind FAB joining the UN-convened Net Zero Banking Alliance as the first bank in the UAE and GCC to make a net-zero commitment by 2050.

Under his leadership, FAB has been recognized as a leading bank within sustainability. Shargiil is recognized by LinkedIn as the Top Voice Green in the Middle East. He is also a member of the Net Zero Banking Alliance Steering Group and UNEP FI (UN Environment Programme – Finance Initiative) Banking Board.

Shargiil holds a Bachelor of Business Administration and an Executive MBA from the Copenhagen Business School in Denmark – and has completed Executive Education at INSEAD in France.

Shargiil lives in Abu Dhabi with his wife, Nina, and their two children.

To my mother, who has always loved, guided and inspired me

Shargiil Bashir

FROM VISION TO IMPACT: IMPLEMENTING SUSTAINABILITY IN YOUR BUSINESS

AUSTIN MACAULEY PUBLISHERS®

LONDON * CAMBRIDGE * NEW YORK * SHARJAH

ISBN 9789948737094 (Paperback)
ISBN 9789948737100 (E-Book)

Application Number: MC-10-01-9179898
Age Classification: E

The age group that matches the content of the books has been classified according to the age classification system issued by the UAE Media Council.

Printer Name: iPrint Global Ltd
Printer Address: Witchford, England

First Published 2024
AUSTIN MACAULEY PUBLISHERS FZE
Sharjah Publishing City
P.O Box [519201]
Sharjah, UAE
www.austinmacauley.ae
+971 655 95 202

I humbly extend my deepest gratitude to the individuals whose support, invaluable guidance, and inspiration were instrumental in the creation of this book.

First and foremost, I express my profound appreciation to my beloved mother, whose unwavering support and invaluable life lessons have shaped me as a person.

To my wife, your unconditional love, endless support, and wise guidance have been instrumental in every step of this journey. I am also grateful for the challenges and encouragement from our son, Ehaan, and the always-supportive gestures from our daughter, Emel.

To my entire family, your love, understanding, and encouragement have been a constant wellspring of strength and motivation.

I am deeply thankful to all my leaders through the years, whose mentorship, wisdom, and teachings have played a pivotal role in my personal and professional growth. Your guidance and insights have led me to evolve in ways beyond my imagination.

To my supportive colleagues, your insights, collaboration, support, and challenges have pushed me to broaden my horizons and think innovatively. Our shared commitment to making a positive impact on the world has been truly inspiring and transformative.

Special thanks to my employers, Danske Bank and First Abu Dhabi Bank, who through the years have supported, developed, and guided me to achieve more.

Lastly, I offer my gratitude to the Almighty for blessing me with opportunities I never thought possible.

To every individual who has been part of this remarkable journey, I extend my heartfelt appreciation. With deepest gratitude and warm regards,

Shargiil Bashir

All royalties made from this book will be donated to NGO's for the betterment of our planet.

Table of Content

Introduction

This book is all about sustainability – what it means, why it's important, and how it can make a difference in the world. It covers everything from the big-picture impact it can have on society and the planet we live on to how companies and employees can adopt sustainability. This book addresses the growing concern worldwide about the negative impact humans have on the environment, using up resources faster than they can be replenished and creating social imbalances. The goal is to offer companies and their employees practical solutions and ideas that can be implemented to help make a positive difference. By understanding the importance of sustainability and taking action, we can help protect the planet for future generations.

What Is Sustainability & Why Is It Important?

Sustainability is a big word that refers to the act of taking care of the environment, economy, and society in a way that does not harm them. It means being able to meet the needs of people today without hurting the ability of future generations to meet their own needs. Sustainability is all about keeping a balance between these three factors so that we can use our resources without using them up completely.

It's very important to think about sustainability because it helps us to keep our planet healthy and our resources intact. When we care for the environment, we can reduce things like pollution and waste, which help preserve biodiversity and keep our planet healthy. When we think about the economy, we can make sure that businesses are doing well while also making sure that people are being treated fairly. When we think about society, we can try to make sure that everyone has access to the things they need to live a good life, like food, water, and shelter.

By working towards sustainability, we can make sure that future generations will have the same opportunities to thrive as we do. It's important to take action now to make sure that our planet and its resources are preserved for many years to come.

Sustainability has become crucial for companies as all their key stakeholders, like customers, employees, shareholders, and regulators, are focusing on sustainability.

Who Is the Audience For This Book?

This book is for anyone who is interested in learning more about sustainability and how it can be applied to different areas of life and work. It's written in a way that is easy to understand and caters to a diverse audience, including students, academics, policymakers, business professionals, and individuals who want to make a positive difference in the world.

By reading this book, you'll get a comprehensive understanding of sustainability. You'll learn about how it can be practically applied in different contexts. The book includes case studies, practical strategies, and expert insights, providing you with everything you need to implement sustainable practices in your business, policymaking, personal life, or community initiatives.

The goal of this book is to inspire action and provide a roadmap for creating a more sustainable future. Whether you're interested in reducing your carbon footprint, promoting social and economic equity, or preserving biodiversity, this book will give you the guidance and tools you need to make a positive impact. By reading this book, you'll be empowered to take action and contribute to a more sustainable world.

It is my sincere hope that this book will inspire others on their journeys toward sustainability and contribute to a better world. Therefore, I am committing that all royalties made from this book will It is my sincere hope that this book will inspire others on their journeys toward sustainability and contribute to a better world. Therefore, I am committing that all royalties made from this book will be donated to NGO's for the betterment of our planet.

Chapter 1
Leadership Commitment

"Sustainability is not a cost; it's an investment in the future. Leaders who understand this will be the ones who thrive in the 21st century."

In the evolution of business, a paradigm shift has emerged—one that transcends profit margins and market dominance. Sustainability was once considered an idealistic pursuit but has become imperative for companies navigating a world fraught with environmental challenges, social inequities, and economic uncertainties.

At the heart of any successful sustainability journey lies a fundamental truth: the commitment of senior leadership. It's not merely a checkbox on a corporate to-do list but a transformative ethos that must permeate the organizational fabric from the top down. Without this staunch support, the path to meaningful change becomes an uphill battle, where efforts risk dissipating in the corridors of middle management or, worse, stagnating altogether.

But why does the allegiance of senior leadership hold such unparalleled significance? At its core, it's about setting the trajectory for the entire enterprise, steering it toward a future where profitability doesn't come at the expense of the planet or society. This commitment is the catalyst for aligning strategy, values, and actions with a broader purpose—to not only thrive in the present but to secure a sustainable legacy for future generations.

Securing this commitment, however, isn't a simple task. It requires a compelling narrative that goes beyond profit margins, emphasizing the symbiotic relationship between sustainable practices and long-term business resilience. It necessitates a vision that fuses ethical responsibility with economic prudence, showcasing how sustainability isn't a burden but an opportunity for innovation, growth, and societal contribution.

The Difference Between ESG & Sustainability

Often, concepts like ESG are used interchangeably with sustainability. While ESG factors are crucial components of sustainability, they represent a subset of a broader sustainable approach. Understanding this distinction is key to aligning leadership focus effectively.

In ESG, **Environmental** refers to factors related to a company's impact on the environment. It includes aspects like carbon emissions, energy efficiency, waste management, and natural resource conservation. **Social** factors encompass a company's relationships and impacts on society. This involves considerations such as labor practices, employee relations, diversity, equity, community engagement, and human rights. Finally, the **Governance** factor focuses on the company's internal structures, policies, and leadership. It includes elements like board independence, executive compensation, transparency, and ethics.

ESG factors are used as a framework for evaluating a company's performance in these areas. Investors, stakeholders, and financial analysts use ESG criteria to assess risks and opportunities associated with a company's operations, looking beyond just financial metrics to evaluate its impact on society and the environment.

Sustainability is a more comprehensive and holistic concept. It encompasses ESG factors but extends beyond them to consider long-term viability and balance between economic, environmental, and social aspects. It's about meeting present needs without compromising the ability of future generations to meet their needs.

Sustainability emphasizes enduring practices that promote responsible use of resources, resilience to environmental changes, social equity, and ethical governance. It's about ensuring that business operations are not only profitable but also contribute positively to society and the planet.

Key Differences:

- **Focus and Scope:** ESG is a set of specific factors used for assessment, whereas sustainability is a broader philosophy encompassing ESG factors and long-term viability.
- **Timeframe:** ESG factors often focus on current performance metrics, while sustainability emphasizes long-term impact and resilience.

- **Purpose:** ESG factors are often used as a tool for analysis by investors and stakeholders, whereas sustainability is a guiding principle for responsible business practices.

In essence, while ESG factors are essential components of sustainability, sustainability encapsulates a larger philosophy that goes beyond immediate metrics and integrates environmental, social, and governance considerations for long-term viability and responsible business conduct.

The Role of the Board and Senior Management in Sustainability

The leaders of the company, including the Board and senior management, are responsible for driving this change towards sustainability. By endorsing the importance of sustainability, they are taking it from being just a task for a particular department to a crucial aspect embedded into the entire organization's culture. This means that everyone in the company must have a mindset that prioritizes sustainability, not just treating it as something that can be put aside. The decisions, directives, and attitudes of these leaders will determine how sustainability is integrated into the core strategy and operations of the company. By making sustainability a part of their culture, the company can achieve long-term success while also being responsible towards the environment and society.

1. **Setting The Vision & Strategy**

The Board plays a crucial role in setting the long-term vision and strategic direction of the company, including sustainability goals. They set the direction and establish the framework within which Senior Management formulates sustainability strategies aligned with the company's purpose.

Unilever stands as a notable example where the Board actively supports sustainability. Their Sustainable Living Plan, initiated by the Board, outlines ambitious goals for reducing environmental impact while improving social conditions.[i]

2. Accountability & Oversight

The Board plays a powerful role in helping guide the company on its sustainability journey.

They are responsible for setting the strategic direction on sustainability, while the Senior Management is responsible for overseeing the integration of sustainability into business practices. They ensure that sustainability initiatives are implemented effectively and in line with the company's values and legal obligations.

Interface Inc., a carpet tile manufacturer, transformed its operations under the guidance of its founder, Ray Anderson. The Board played a crucial role in adopting and monitoring sustainability targets, focusing on reducing waste and carbon emissions.[ii]

3. Risk Management & Opportunity Identification

Identifying risks associated with sustainability issues such as climate change, supply chain vulnerabilities, and regulatory changes is a critical aspect. Boards and Senior Management assess these risks and seek opportunities that align with sustainability goals.

The Board and senior management should assess these risks for several reasons:

- Financial impact: Climate change and related regulations can impact the business strategy or product offering in the short or long term. This can have significant financial implications for companies.
- Reputational risk: Companies that are perceived as not taking climate change seriously can suffer reputational damage, which can have long-term negative effects on their brand and ability to attract customers, investors, and talent.
- Legal and regulatory compliance: As governments around the world implement more stringent climate-related regulations, companies need to ensure they are in compliance to avoid potential legal and financial penalties.

Tesla, led by CEO Elon Musk and its Board, recognized the environmental risks associated with traditional automobiles. They directed the company's focus towards sustainable transportation by producing electric vehicles, addressing climate-related risks.[iii]

4. Transparency & Reporting

Boards and Senior Management are accountable for transparently reporting to stakeholders on the company's sustainability performance. They ensure that meaningful metrics and disclosures are communicated, fostering trust and credibility.

Danone, a multinational food-products corporation, aligns its reporting with global sustainability frameworks. The Board ensures comprehensive disclosures, including environmental impact and social responsibility, enhancing transparency.[iv]

5. Cultural Influence & Leadership

Leadership is instrumental in shaping the organizational culture towards sustainability. Their actions and decisions demonstrate a commitment that permeates throughout the company, fostering a culture of responsibility and innovation.

Patagonia, a clothing company, is renowned for its sustainability ethos. The Board and Senior Management actively endorse sustainability practices, reflecting a commitment that resonates across the organization.[v]

In general, the Board and Senior Management aren't merely observers; they are drivers of sustainability. Their commitment, strategic decisions, and alignment with sustainable values are integral to embedding sustainability into an organization's DNA, transforming it into a responsible global citizen.

Importance of Commitment from Leadership

The commitment from leadership is the cornerstone of any successful sustainability initiative within an organization.

Leadership commitment establishes the overarching narrative and direction for sustainability efforts. When leaders prioritize sustainability, it signals to the entire organization that these initiatives are not peripheral but central to the company's ethos and strategy.

Secondly, leadership shapes organizational culture. A commitment to sustainability from the top fosters a culture where environmentally and socially responsible practices are valued and integrated into everyday operations. It encourages innovation and employee engagement around sustainability.

Senior leadership's commitment ensures alignment between sustainability goals and the overall business strategy. It directs resources, influences decision-making, and aligns various departments towards common sustainability objectives.

Leadership committed to sustainability recognizes the importance of long-term value creation over short-term gains. Sustainable practices often yield economic benefits in the long run, reducing risks, enhancing brand reputation, and ensuring resilience in the face of environmental and social challenges.

Moreover, commitment to sustainability builds trust among stakeholders—employees, customers, investors, and communities. It demonstrates authenticity and a genuine commitment beyond profit, enhancing the company's credibility and reputation.

Leadership commitment to sustainability helps identify and mitigate risks associated with environmental, social, and governance factors. It ensures the company is better equipped to handle changes in regulations, market demands, and environmental impacts.

Sustainability commitment fosters a culture of innovation, encouraging the development of new technologies, processes, and business models that are more resource-efficient, environmentally friendly, and socially responsible.

In a world facing pressing environmental and social challenges, leadership commitment to sustainability demonstrates corporate responsibility and contributes to addressing these global issues.

IKEA has showcased its leadership commitment to sustainability by integrating it into its core strategy. Through leadership-driven initiatives like sustainable sourcing, energy efficiency, and waste reduction, IKEA aligns its business model with sustainability goals, showing how leadership commitment drives impactful change.[vi]

In essence, commitment from leadership is the catalyst that propels an organization towards a sustainable future. It influences strategy, culture, and decision-making, steering the company towards responsible and impactful practices that benefit both the business and society as a whole.

How To Get Buy-In From Leadership?

Getting buy-in from leadership for sustainability initiatives can be a nuanced process. Here are some ways to secure their commitment:

1.   Link Sustainability to Business Goals & Strategy

Connect sustainability initiatives directly to the organization's overarching goals and strategy, showing how sustainability can enhance profitability, reduce risks, and foster innovation. Present a clear business case highlighting potential cost savings, market opportunities, and competitive advantages.

2.   Educate & Raise Awareness

Provide comprehensive education and awareness sessions to leaders. Share industry trends, success stories, and data-driven insights that demonstrate the relevance and impact of sustainability on the company's bottom line and reputation.

3.   Highlight Risks & Opportunities

Illustrate the risks associated with ignoring sustainability trends, such as regulatory changes, supply chain disruptions, or reputational risks. Simultaneously, emphasize the opportunities that sustainability presents—new markets, increased customer loyalty, and operational efficiencies.

4.   Demonstrate Short-Term Wins

Start with smaller, achievable sustainability projects that showcase immediate benefits. Highlighting these wins can build momentum and credibility, making it easier for leaders to see the value in broader, long-term sustainability initiatives.

5. Secure Leadership Champions

Identify influential advocates within the leadership team who can champion sustainability initiatives. Their support can sway opinion and help bridge the gap between sustainability advocates and more skeptical leaders.

6. Align With Organizational Values

Illustrate how sustainability aligns with the company's core values and mission. Emphasize the ethical and moral imperative of responsible business practices, resonating with the leadership's personal and professional ethos.

7. Use Data & Metrics

Present data-backed arguments and metrics that quantify the impact of sustainability initiatives. Tangible figures showcasing cost savings, reduced resource use, or improved brand perception can be persuasive.

8. Foster Dialogue & Collaboration

Encourage open discussions and collaboration between leaders and sustainability advocates. Create forums where leaders can voice concerns, ask questions, and participate in decision-making regarding sustainability strategies.

Interface Inc. successfully gained leadership buy-in for sustainability when its CEO, Ray Anderson, became an ardent advocate for environmental stewardship. His commitment led to significant transformations within the company, showcasing how a passionate leader can drive change.

Securing buy-in from leadership is about building a compelling narrative that speaks to both the business imperatives and the moral obligation of sustainability. It requires a strategic approach that emphasizes alignment with core business goals while showcasing the long-term benefits and positive impacts of sustainability initiatives.

Tone From the Top Is the Key

Tone from the top is not merely a phrase; it's a beacon guiding organizational change. It's about leaders not just speaking about sustainability but embodying it in their decisions and actions. Authenticity in their commitment resonates throughout the ranks, fostering a culture where sustainability isn't an added responsibility but an inherent part of doing business.

The behavior and attitudes of leaders set the cultural norms within an organization. When leaders embody the values and practices they expect from their teams, it establishes a precedent for everyone else to follow.

Leaders serve as role models. Their actions speak volumes and carry significant weight. When they actively engage in and prioritize sustainability efforts, it signals to employees that these initiatives are not optional but essential aspects of the company's identity.

Leadership's commitment to sustainability ensures that it is not perceived as a side project but as a core component of the company's strategy. This alignment emphasizes that sustainability is integral to achieving the organization's overall objectives.

The "tone from the top" guides decision-making processes throughout the organization. When sustainability is embedded in leadership discussions and considerations, it influences strategic choices, resource allocations, and investments.

Leadership commitment to sustainability fosters a culture of accountability. It encourages individuals at all levels to take responsibility for integrating sustainability into their roles and activities.

Consistent demonstration of commitment to sustainability by top leadership builds trust and credibility, both internally and externally. This enhances the organization's reputation and stakeholder relationships.

Patagonia stands as an exemplary case where the "tone from the top" is integral to their sustainability ethos. The company's founder, Yvon Chouinard, instilled a commitment to environmental sustainability. This ethos is pervasive throughout the organization, influencing product design, sourcing, and advocacy efforts.

In short, the "tone from the top" shapes the organizational mindset and behavior. It's not just about verbal support but about leaders embodying sustainability in their actions and decisions. Their commitment becomes the driving force that permeates every level of the organization, fostering a genuine culture of sustainability.

ESG-Related Responsibilities Within an Organization

Responsibilities related to ESG within an organization are distributed across different levels: strategic, tactical, and operational. Each level carries distinct roles in driving sustainable initiatives and embedding them into the organizational framework.

1. **Strategic Level:**

At the strategic level, the responsibility lies in setting the overall direction and vision for ESG integration within the organization. This involves aligning ESG goals with the company's mission and long-term strategy.

The Board of Directors and senior leadership play a crucial role in defining the strategic priorities related to ESG. They set policies, establish frameworks, and ensure that ESG considerations are integrated into the company's strategic planning.

Identifying ESG risks and opportunities is a strategic responsibility. It involves conducting risk assessments related to environmental impact, social issues, and governance structures to inform long-term planning.

2. **Tactical Level:**

Translating the strategic vision into actionable plans falls under tactical responsibilities. This involves developing specific initiatives and programs to meet the ESG goals set by leadership.

Tactical responsibilities involve integrating ESG considerations into various departments. For instance, the HR department might focus on diversity and inclusion initiatives, while operations might concentrate on energy efficiency.

Formulating strategies to engage stakeholders—such as investors, employees, customers, and communities—is a tactical responsibility. This includes communication plans and engagement initiatives related to ESG matters.

3. Operational Level:

Operational responsibilities involve the actual execution of ESG initiatives on a day-to-day basis. It includes implementing sustainable practices, managing resources efficiently, and ensuring compliance with ESG standards.

Collecting relevant ESG data and generating reports to track progress and performance against set goals is an operational responsibility. This involves measuring and reporting on environmental impact, social initiatives, and governance practices.

Ensuring that employees are trained and equipped to understand and contribute to ESG initiatives is an operational duty. It involves embedding sustainability practices into the daily operations and decision-making processes.

These responsibilities can be across multiple departments or by establishing a dedicated ESG department within an organization. A dedicated department can play a pivotal role in integrating ESG considerations into the organization's overall strategy, operations, and decision-making processes.

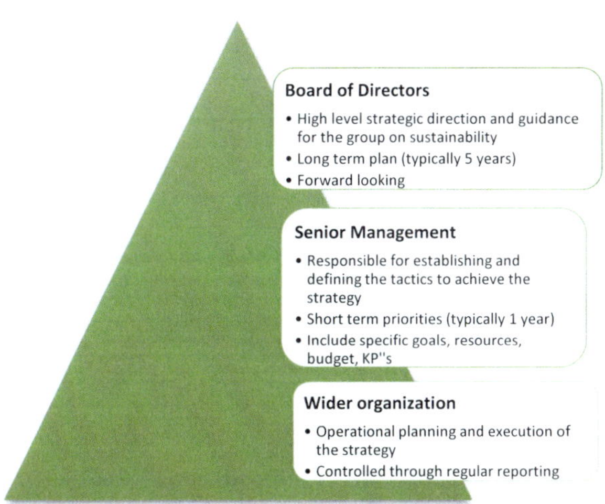

Nestlé exemplifies the distribution of ESG res
ponsibilities within an organization. Their Board sets clear sustainability goals, the management devises strategies for achieving these goals, and operations focus on implementing measures like sustainable sourcing, reducing water usage, and improving social impact.

In short, responsibilities for ESG at different levels ensure that sustainability is not just a theoretical concept but is integrated into an organization's strategic planning, tactical decision-making, and daily operations, fostering a holistic approach towards sustainability and responsible business practices.[vii]

Stakeholder Analysis

The most compelling reason for leadership commitment lies in the impact on stakeholders. Employees, customers, shareholders, and regulators—they all play interconnected roles in an organization's ecosystem. Sustainability isn't a concern limited to environmentalists; it's a universal language that resonates with each stakeholder group. It embodies ethical practices, responsible governance, and a commitment to the greater good.

1. Employees:

Sustainable practices can enhance employee morale and engagement. Employees often feel more connected to an organization that aligns with their values, fostering a sense of purpose and pride in their work.

Sustainability initiatives can create a healthier, safer, and more inclusive workplace. Employees appreciate initiatives that prioritize their well-being and support a positive work environment.

Many individuals, especially the younger workforce, seek out employers with strong sustainability commitments. Companies demonstrating environmental and social responsibility can attract top talent.

Among employed adults surveyed by **Deloitte Consumer Center** in March 2023 for its global State of the Consumer survey, 69% said they want their companies to invest in sustainability efforts, including reducing carbon, using renewable energy, and reducing waste. [viii]

2. Customers:

Sustainable practices can enhance brand reputation and customer loyalty. Consumers increasingly favor companies that demonstrate a commitment to environmental conservation, ethical sourcing, and social responsibility.

Sustainability can influence purchasing decisions. Consumers are more inclined to choose products and services from companies that prioritize sustainability, driving market demand for eco-friendly and ethically produced goods.

Communicating sustainability efforts builds trust. Transparent reporting about sustainable practices fosters credibility and strengthens the relationship between companies and their customers.

According to a study by **Bain & Co**, consumers say their environmental concerns are increasing due to extreme weather, and they are willing to change their behavior and pay 12% more for sustainable products.[ix]

3.   Shareholders & Investors:

Sustainability practices can mitigate risks associated with environmental, social, and governance issues. Investors seek companies with robust sustainability strategies to ensure long-term resilience.

Companies with strong ESG performance often demonstrate better financial performance over the long term. Investors recognize the correlation between sustainability and profitability.

Anticipating and adhering to evolving sustainability regulations ensures compliance and reduces legal and reputational risks, which is crucial for shareholders.

ESG assets will hit $50 trillion by 2025, representing more than a third of the projected $140.5 trillion in total global assets under management, according to Bloomberg.[x]

According to a 2022 study by **Capital Group**, 89 percent of investors consider ESG issues in some form as part of their investment approach. In addition, only 13 percent of global investors see ESG as a "passing fad that will eventually go out of fashion".[xi]

At the annual general meeting (AGM) of UK bank **Barclays**, a coalition of investors represented by responsible investing NGO ShareAction pushed the bank to take stronger actions on ESG issues. These investors included notable institutions such as AkademikerPension, Barrow Cadbury Trust, Brunel Pension Partnership, Candriam, Ethos Foundation, and others. The primary focus of their advocacy was urging Barclays to cease financing new oil and gas projects and to provide more transparency regarding its assessment of clients' climate transition plans.

This shareholder action is part of a broader campaign coordinated by ShareAction, which started in February and targeted several major banks, including Crédit Agricole, Deutsche Bank, and Societe Generale, to commit to ending financing for new oil and gas fields. The campaign emphasizes the importance of aligning banking activities with climate goals, particularly in light of commitments made by other European banks towards achieving net-zero emissions by 2050.

Barclays, in response to the campaign and shareholder pressure, announced policies aimed at limiting its financing for emissions-intensive energy sectors. However, ShareAction noted that these policies fell short of the minimum standards of ambition required to address climate concerns adequately.

During the AGM, shareholders pressed Barclays' Chair Nigel Higgins to commit to halting financing for new oil and gas fields and to disclose detailed information about its evaluation of clients' climate transition plans. While Higgins acknowledged the ongoing work on transition plans and the engagement with ShareAction, he also emphasized the bank's responsibility to work with energy companies to meet energy demand while transitioning to a more sustainable future.[xii]

This example illustrates how shareholders are leveraging their influence at AGMs to drive corporate accountability and action on ESG issues, particularly regarding climate change and sustainable financing practices. It showcases the growing importance of ESG considerations in investment decisions and corporate governance, highlighting the role of investors in promoting sustainability within the financial sector.

4. Regulators & Authorities:

Regulators are increasingly focused on sustainability due to global concerns about climate change, resource depletion, and social equity. Sustainable practices align with regulatory initiatives aimed at addressing these issues.

Regulators often establish frameworks and standards for ESG reporting. Compliance with these standards ensures transparency and consistency in reporting environmental and social impacts.

Regulators represent the broader community's interests, including environmental protection, social welfare, and ethical business practices. Sustainability aligns with these public interests.

According to **ESG Book**, a global leader in sustainability data and technology, global ESG regulations have increased by 155 per cent over the past decade as the rapid growth of sustainability-based policy interventions continues to shape financial markets.

According to research by ESG Book, 1,255 ESG regulations have been introduced worldwide since 2011, compared to 493 regulations published between 2001 and 2010. Since the turn of the millennium to the present day, there has been a 647 per cent increase in ESG regulations.xiii

Unilever illustrates the impact of sustainability on stakeholders. Their Sustainable Living Plan not only drives employee engagement and customer loyalty but also attracts investors interested in the company's commitment to sustainable growth.

In short, sustainability matters to stakeholders because it addresses their concerns, values, and expectations. Embracing sustainability isn't just a corporate responsibility; it's a means to foster positive relationships, mitigate risks, and create long-term value for all stakeholders involved.

Leadership commitment isn't just a prerequisite; it's the cornerstone on which the edifice of sustainability stands. Without this steadfast dedication from the top, the journey toward a more sustainable future remains an aspiration rather than a tangible reality.

Conclusion

Sustainability is no longer a choice but a necessity, and its integration into business practices is imperative for organizations navigating a world challenged by environmental, social, and economic uncertainties.

Sustainability, often encapsulated by the triple bottom line approach of "People, Planet, and Profit," embodies the interconnectedness of environmental stewardship, social responsibility, and economic viability. Achieving sustainability requires striking a delicate balance among these three pillars, ensuring that initiatives are not only environmentally and socially responsible but also economically feasible. By prioritizing people, organizations commit to fostering inclusive workplaces, supporting employee well-being, and engaging with local communities. Planet-focused efforts center on minimizing environmental impacts, conserving natural resources, and mitigating climate change through sustainable practices. Profit considerations involve generating financial returns while simultaneously creating value for society and minimizing negative externalities. Finding alignment and synergy among these objectives is essential for achieving sustainable business success. It requires thoughtful decision-making, innovative solutions, and a commitment to long-term prosperity for both the organization and the broader ecosystem in which it operates. Through strategic planning and collaboration, organizations can navigate the complex landscape of sustainability, driving positive outcomes for people, the planet, and profit alike.

The core of any successful sustainability journey lies in senior leadership's commitment. This isn't a token gesture but a transformative ethos that must resonate through every facet of the organization, from the highest echelons of leadership to the day-to-day operations. Without this unwavering support, efforts towards sustainability risk being confined to rhetoric or mid-level management, lacking the impetus needed for real change.

Why is the allegiance of senior leadership so pivotal? At its essence, it's about setting the course for the entire enterprise, steering it towards a future where profitability is symbiotic with environmental and societal well-being. This commitment is what underpins a business's strategy, values, and actions, ensuring not just survival in the present but a legacy of sustainability for future generations.

But securing this commitment isn't straightforward. It demands a compelling narrative beyond profit margins, highlighting the symbiotic relationship between sustainability and long-term business resilience. It necessitates a vision that unifies ethical responsibility and economic prudence, demonstrating how sustainability isn't a burden but an opportunity for innovation, growth, and societal contribution. Embracing ESG is a transformational journey in leadership, be it at the board level or within senior management, and holds the mantle for driving this transformation and change towards sustainability. They are tasked with aligning the organization's strategy, values, and actions with its purpose—a purpose that's rooted in environmental stewardship, social equity, and ethical governance.

Boards play a critical role in setting the strategic direction for sustainability, ensuring that goals are comprehensive, visionary, and integrated into the organization's mission. They provide oversight, guide risk management strategies, and set transparency standards for reporting. Meanwhile, senior management ensures that these strategic goals are translated into tangible, day-to-day actions that embed sustainability into the organization's culture and operations.

Moreover, leadership fosters a culture where sustainability isn't seen as an added responsibility but an intrinsic part of doing business. They cultivate an environment where employees feel empowered to contribute to sustainability initiatives, where innovation flourishes, and where the company's reputation is bolstered by its commitment to environmental and social responsibility.

Commitment from the top isn't just about verbal assurances; it's about embodying sustainability in actions and decisions. Authenticity in leadership's commitment is what resonates through the ranks, ensuring that sustainability isn't just a peripheral agenda but a core aspect of the organization's identity.

Securing buy-in from leadership for sustainability initiatives involves a strategic approach. It's about presenting a clear business case, highlighting short-term wins, aligning with organizational values, and fostering dialogue and collaboration. Leadership champions—those who embody sustainability in their ethos and actions—are instrumental in swaying opinions and bridging the gap between sustainability advocates and skeptical leaders.

Leadership commitment isn't just about the present; it's about setting the course for the future. By prioritizing sustainability today, leaders ensure that their organizations not only survive but thrive in the 21st century—a century that demands not just profit but purpose, not just growth but sustainability.

Chapter 2
Assess Current State

"Progress is impossible without change, and those who cannot change their minds cannot change anything." – George Bernard Shaw.

Assessing the current state of a company's impact—both positive and negative—on its surroundings is the foundational step towards embracing sustainability. It's about gaining a comprehensive understanding of how operations, practices, and decisions influence the environment, society, and the economy.

Imagine setting sail without a map; without understanding your starting point, addressing challenges and charting a course towards sustainability becomes a daunting task. It's important to take a close look at your current state; this helps you figure out which direction you want to go and what actions you need to take to get there.

Without a clear understanding of your current impact, it's akin to navigating through uncharted waters blindfolded. You might sense the vastness and potential hurdles, but the specifics elude you. This assessment discloses the landscape, shedding light on areas that demand attention. It transforms ambiguity into a roadmap, providing a clear direction on where to focus your sustainability efforts.

The magnitude of change required for a sustainable transformation becomes apparent when you know your current state. It's about realizing how little tweaks or monumental shifts can lead to positive impacts. Whether it's optimizing resource usage, reducing emissions, or redefining supply chain practices, understanding your starting point helps quantify the scale of change needed.

What is deemed good or bad is often relative and evolving. By assessing your current state, you establish a benchmark. This benchmark becomes the yardstick against which you measure the positive and negative aspects of your impact. It's not a static judgement but a dynamic understanding, providing context for what needs improvement and what can be celebrated.

Assessing your current state isn't just about numbers; it's about maturity. It's a journey into the organizational psyche, uncovering how deeply sustainability is ingrained. Through interviews, surveys, and rigorous analysis, you gain insights into the organization's maturity on the topic. Are sustainability practices merely superficial, or are they deeply woven into the fabric of decision-making and daily operations?

Armed with a comprehensive understanding of your current impact, you gain insights into pathways for improvement. It's not just about identifying problems; it's about envisioning solutions. Interviews and analyzes reveal the untapped potential within your organization, guiding you towards initiatives that align with your values and contribute to a sustainable future.

Acknowledging your current state establishes a baseline for accountability. It's not a blame game but an acknowledgement that, from this point forward, every action is intentional. This understanding fosters a culture of accountability where employees, stakeholders, and leadership collectively strive for continuous improvement.

In essence, assessing your current state is not a static exercise but a dynamic process that illuminates the path forward. It's the tool that transforms uncertainties into opportunities, challenges into solutions, and aspirations into achievable goals. With this understanding, you step into the world of sustainability with clarity, purpose, and a roadmap tailored to your unique organizational landscape.

Data Gathering

Data gathering related to ESG is crucial as it forms the foundation for informed decision-making, transparency, and accountability in businesses. This data allows companies to measure their performance across ESG metrics, identify risks, enhance investor relations, create long-term value, and drive continuous improvement. It enables companies to set benchmarks, mitigate risks, attract investment, comply with regulations, and align business strategies with sustainability goals. Overall, ESG data is integral to sustainable growth, resilience, and responsible business practices in today's corporate environment.

The challenges related to the lack of data in ESG, particularly in the environmental aspect, present significant hurdles for companies aiming to integrate ESG into their strategies effectively. Historically, companies have not gathered or disclosed enough internal data related to ESG, leading to gaps in understanding their environmental impact. Additionally, the absence of national or regional institutions gathering such data further compounds this challenge.

One of the primary obstacles is identifying reliable data sources. Companies need to determine where to source ESG data, whether internally (e.g., diversity metrics) or externally (e.g., industry benchmarks). Ensuring ongoing maintenance and relevance of ESG data is another hurdle, requiring alignment across business units and robust security measures.

Furthermore, shifting operations and culture to incorporate ESG strategies is complex, requiring buy-in from leadership down to mid-level managers. Implementing ESG technology can facilitate this transition but requires overcoming resistance and fostering ownership of ESG goals.

The lack of standardized frameworks and reporting requirements in the field of ESG presents a significant challenge for organizations. Without clear guidelines and uniform metrics, companies must navigate a complex landscape of evolving standards and anticipate future regulatory changes. This uncertainty can make it challenging to establish consistent reporting practices and compare performance across different organizations.

Moreover, monitoring stakeholder sentiment and managing third-party ESG risks add another layer of complexity. Stakeholders, including investors, customers, and communities, have varying expectations and priorities regarding ESG practices. Keeping abreast of these expectations and effectively communicating ESG efforts to stakeholders requires robust data collection, analysis, and reporting mechanisms.

Additionally, managing ESG risks associated with third-party relationships, such as suppliers and business partners, is crucial. These relationships can significantly impact an organization's ESG performance and reputation. Ensuring transparency and accountability throughout the supply chain requires comprehensive data solutions and effective risk management strategies.

Overall, addressing the challenges related to the lack of standardized frameworks, evolving standards, stakeholder expectations, and third-party ESG risks underscores the critical importance of reliable and comprehensive data solutions in driving ESG progress and creating long-term value for organizations.[xiv]

1. Environmental Impact Data Gathering:

GRI (Global Reporting Initiative) & GHG Protocol (Greenhouse Gas Protocol):

GRI Standards[1]: Utilize GRI guidelines for reporting to gather comprehensive environmental data. These standards offer a framework for reporting on various environmental impacts such as energy consumption, water usage, emissions, and waste generation. **GHG Protocol[2]:** The GHG Protocol provides methodologies for measuring and managing greenhouse gas emissions. It helps quantify emissions across scopes, enabling a thorough assessment of the company's carbon footprint

Environmental Footprint Assessment:

- **Energy Consumption:** Collect data on energy usage across facilities, operations, and transportation. Determine electricity consumption, fuel usage, and other energy sources utilized by the company.
- **Carbon Emissions:** Quantify carbon emissions across scopes, including direct (Scope 1) and indirect (Scope 2 and Scope 3) emissions. This involves tracking emissions from owned facilities, purchased electricity, and supply chain activities.
- **Water Usage:** Gather data on water consumption and wastewater generation. Assess water usage efficiency and identify opportunities for conservation.

2. Social Impact Data Collection:

- **Employee Data:** Gather information on workforce diversity, employee turnover rates, health safety records, and training programs.

---

[1] www.globalreporting.org/standards
[2] www.ghgprotocol.org/

- **Community Engagement:** Collect data on community projects, philanthropic initiatives, and the impact of these efforts on local communities.
- **Supplier Relationships:** Assess supplier performance, compliance with labor standards, and adherence to ethical sourcing practices.

3. Governance Model Assessment:

- **Board Structure:** Analyze the composition and independence of the board of directors. Evaluate board diversity, expertise, and its influence on decision-making.
- **Ethical Practices:** Gather data on the company's policies and practices related to ethical conduct, compliance, anti-corruption measures, and transparency.
- **Shareholder Rights:** Assess mechanisms for protecting shareholder rights, shareholder engagement practices, and governance structures supporting stakeholder interests.

Toyota uses the GRI Sustainability Reporting Standards and the GHG Protocol to report on its environmental impacts comprehensively. The company assesses its energy consumption, emissions, and water usage to track progress and set targets for sustainability.[xv,xvi]

Toyota has been working on various initiatives to reduce its environmental impact. The company is committed to reducing CO_2 emissions and achieving carbon neutrality globally by 2050 for its vehicles and operations. In Europe, Toyota aims to achieve full carbon neutrality by 2040. Toyota is also improving water usage, promoting end-of-life and recycling technologies, and establishing a society in harmony with nature.[xvii]

Toyota has set an ambitious goal to reduce the negative environmental impacts of manufacturing and driving vehicles as close to zero as possible and to make net positive impacts on society. The company is promoting measures based on six challenges, including new vehicles with zero CO_2 emissions, manufacturing plants with zero CO_2 emissions, life cycle zero CO_2 emissions, minimizing and optimizing water usage, establishing a recycling-based society, and establishing a future society in harmony with nature.[xviii]

Benefits Of Data Gathering:

- **Informed Decision-Making:** Data-driven insights enable informed decisions for setting targets, formulating strategies, and prioritizing actions for improvement.
- **Transparency & Credibility:** Transparent reporting based on reliable data enhances the company's credibility and trust among stakeholders.

Continuous Improvement: Regular data collection facilitates continuous monitoring, allowing for adjustments and improvements in ESG performance over time.

Market analysis from ESG perspective

1. Competition Analysis:

Analyzing the competitive landscape from an ESG perspective involves understanding how companies within the sector are integrating sustainability into their operations. This includes examining their sustainability reports, initiatives, products, services and the degree to which ESG factors are embedded in their strategies.

This is a template that can be expanded and used to create an overview of your competitors. You can fill it out quantitatively, if possible, or qualitatively. The below items are non-exhaustive and can be expanded based on industry:

	Your Company	Competitor 1	Competitor X
ESG Commitments			
ESG Value Proposition			
ESG related products			

Environmental topics, e.g. Emissions, waste, water, use of renewable energy			
Social topics e.g. Diversity, equality, community engagement			
Governance topics, e.g. structure, frameworks implemented			

2. Sustainability Trends:

Identifying prevailing sustainability trends in the sector helps anticipate future market directions. This involves recognizing emerging practices, innovations, and shifts in consumer preferences towards more sustainable products or services. Assess the sustainability trends and directions both in the short and long term. Example of a sustainability trend is sustainable transportation, where the transportation sector is embracing sustainability through the development of electric vehicles, improved public transportation systems, and initiatives to reduce emissions from planes, trains, and automobiles. Also, the fashion industry is increasingly focusing on sustainability by using eco-friendly materials, reducing water and energy consumption in production processes, and promoting ethical labor practices. Consumers are becoming more conscious of the environmental impact of fast fashion and are supporting sustainable and ethical brands.

3. Regulatory Environment:

Understanding the regulatory landscape and how it impacts the sector is crucial. Analyzing existing and potential regulations related to environmental conservation, social responsibility, and governance can help anticipate compliance requirements and shape strategies accordingly.

4. Business Opportunities Linked to Sustainability:

Identifying business opportunities arising from sustainability trends is pivotal. This includes recognizing untapped markets, product/service innovation, cost-saving measures through efficiency improvements, and potential collaborations aligned with sustainability goals.

5. Future Outlook:

Forecasting the sector's future trajectory from an ESG perspective involves analyzing how sustainability factors will influence market dynamics, consumer behavior, and investor sentiment. This outlook aids in shaping long-term strategies and positioning within the evolving market.

6. SWOT For ESG In the Sector:

- **Strengths:** Recognizing the sector's strengths in terms of existing sustainable practices, established brand value, or technological advancements contributing to ESG goals.
- **Weaknesses:** Identifying areas where the sector lags in sustainability, such as high carbon emissions, supply chain vulnerabilities, or social impact concerns.

- **Opportunities:** Highlighting opportunities to capitalize on emerging sustainability trends, technological advancements, or consumer demands for more ethical and sustainable products/services.
- **Threats:** Understanding potential risks posed by factors like regulatory changes, market competition, shifting consumer preferences, or reputational risks due to inadequate ESG practices.

For example, The Renewable Energy Sector is witnessing a robust competition analysis that emphasizes technological advancements in renewable energy sources. Market trends indicate an increased consumer demand for cleaner energy, while regulatory shifts favor renewable energy investments. Opportunities exist in innovations like energy storage solutions, which can help address the intermittency of renewable energy sources and provide a reliable source of power.[xix]

However, threats loom in uncertain policy frameworks and competition from conventional energy sources. Carefully designed policy frameworks, customized to support technologies at differing stages of maturity, will deliver a strong portfolio of renewable energy technologies.[xx]

In short, a comprehensive sectoral analysis from an ESG perspective not only identifies risks and opportunities but also guides strategic decision-making, fostering a proactive approach towards sustainability-driven growth.

Assessing the current of a company's ESG impact.

An "as-is" analysis of a company's ESG impact involves a comprehensive examination of its current performance and practices in the following key areas.

1. Environmental Impact:

According to the Greenhouse Gas Protocol Initiative, there are three scopes of emissions that organizations can report on:

Scope 1: Direct emissions from sources that are owned or controlled by the reporting organization. Examples include emissions from combustion in owned or controlled boilers, furnaces, vehicles, etc.

Scope 2: Indirect emissions from the generation of purchased electricity, steam, heating, and cooling consumed by the reporting organization.

Scope 3: All other indirect emissions that occur in a company's value chain. Examples include emissions from the production of purchased materials, transportation of goods, waste disposal, etc.

The **GHG Protocol Initiative** provides detailed guidance on how to measure and report emissions for each of these scopes.[xxi]

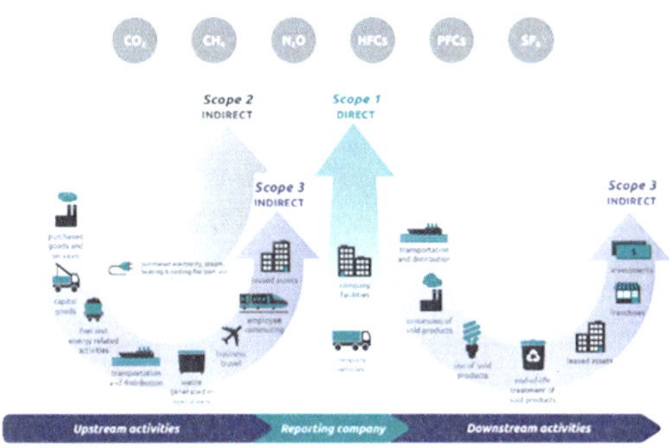

- Calculate your scope 1, 2 and 3 based on GHG protocol, including:

  - **Energy Consumption:** Assess the company's energy usage patterns across operations, identifying areas of high consumption and potential efficiency improvements. This includes electricity, heating, cooling, and transportation.
  - **Resource Usage:** Analyze the utilization of raw materials, water, and other resources. Evaluate waste generation, recycling efforts, and the company's overall resource management practices.
  - **Emissions & Pollution:** Examine the company's carbon footprint, emissions of pollutants, and waste disposal methods. Assess strategies in place for emission reduction and pollution control.

2. Social Impact:

- **Employee Well-being:** Evaluate employee health and safety measures, diversity, equity, and inclusion initiatives, and the overall workplace environment.
- **Community Engagement:** Assess the company's interaction and impact on local communities. This includes philanthropic activities,

community development projects, and efforts to support local economies.

- **Supplier & Partner Relationships:** Examine the company's relationships with suppliers and partners, focusing on fair labor practices, ethical sourcing, and compliance with social standards.

3. Governance Impact:

- **Board Composition:** Analyze the composition and independence of the company's board. Assess the diversity, expertise, and ethical standards of board members.
- **Ethical Business Practices:** Evaluate the company's policies and practices related to transparency, ethics, anti-corruption measures, and compliance with legal requirements.
- **Shareholder Rights:** Assess the protection of shareholder rights, accountability structures, and mechanisms for stakeholders' involvement in decision-making processes.

Apple Inc. has conducted an extensive analysis of its ESG impact, focusing on its supply chain practices. The assessment revealed opportunities for improvements in labor conditions, responsible sourcing of minerals, and reducing environmental impacts across its production chain.[xxii]

Apple has been working on various initiatives to improve air quality, such as reducing carbon emissions at its data centers and investing in renewable energy. The company is also using technology to minimize the environmental impact of its supply chain, packaging, and product transportation.[xxiii] Apple has reduced its emissions by 40 percent since 2015, largely through improvements in energy efficiency, low-carbon design, becoming carbon neutral for corporate operations, and transitioning its supply chain to renewable electricity. [xxii, xxiii]

Apple has set an ambitious goal to become carbon neutral across its entire global supply chain and the life cycle of every product by 2030. The company has been carbon neutral for its global corporate operations since 2020 and is laser-focused on achieving its 2030 goal. Apple has called on its global supply chain to take new steps to address its greenhouse gas emissions and take a comprehensive approach to decarbonization. The company will evaluate the work of its major manufacturing partners to decarbonize their Apple-related operations—including running on 100 percent renewable electricity—and will track yearly progress.[xxii]

Benefits of the As-Is Analysis:

- **Identification Of Strengths & Weaknesses:** It helps identify areas where the company excels and areas that require improvement in terms of ESG performance.
- **Basis For Goal Setting:** The analysis serves as a foundation for setting realistic and impactful ESG goals aligned with the company's current performance.
- **Enhanced Transparency & Reporting:** It enables the company to communicate transparently to stakeholders about its current state, progress, and future commitments.
- **Strategic Decision-Making:** Insights from the analysis inform strategic decisions to prioritize actions that contribute most significantly to sustainable practices and business growth.

Materiality Assessment

Conducting a materiality assessment is pivotal in identifying and prioritizing ESG issues that are most relevant and impactful to an organization and its stakeholders. This process involves evaluating the significance and relevance of various ESG factors to guide resource allocation and strategic focus.

1. Materiality Assessment Process:

Stakeholder Engagement

- Engage with stakeholders—employees, customers, investors, suppliers, communities—to understand their perspectives on ESG issues. Their input helps identify concerns and expectations related to sustainability.

Identifying ESG Factors

- Compile a comprehensive list of potential ESG factors relevant to the industry, business operations, and stakeholder interests. These factors could be (non-exhaustive):

Environmental Factors:

- Carbon emissions and climate change
- Energy efficiency and renewable energy use
- Water usage and management
- Waste management and recycling
- Pollution prevention
- Biodiversity conservation
- Environmental compliance and regulations

Social Factors:

- Labor practices and human rights
- Occupational health and safety
- Diversity and inclusion
- Employee training and development
- Community relations and engagement
- Product quality and safety
- Human capital management

Governance Factors:

- Board diversity and independence
- Executive compensation and incentives
- Anti-corruption policies and practices
- Shareholder rights and engagement
- Ethics and business conduct
- Risk management and oversight
- Data privacy and security

These are just some examples of the many ESG factors that could be considered in a materiality assessment. It's important for companies to tailor their assessment to their specific industry, business model, and stakeholder expectations. Conducting a robust materiality assessment helps companies identify the ESG issues that are most relevant to their business and stakeholders, allowing them to prioritize and focus their efforts on the most significant areas for improvement.

Impact vs. Importance Evaluation

- Assess the significance of each ESG factor based on its impact on the organization and its stakeholders. Consider the direct impact on operations, financial implications, reputational risks, and stakeholder expectations.
- Evaluate the importance of each factor concerning stakeholder interest and perception. Determine the level of concern or expectation stakeholders have regarding these ESG issues.

Prioritization & Mapping

- Plot the identified ESG factors on a materiality matrix, considering their impact and importance. This matrix helps visualize the critical issues that warrant immediate attention (high impact, high importance).

2. Benefits of Materiality Assessment:

Focused Strategy Development

- Helps in crafting a focused and targeted sustainability strategy by highlighting key areas that require attention and improvement.

Resource Allocation:

- Guides efficient allocation of resources by prioritizing efforts towards addressing issues with the most significant impact on stakeholders and the organization.

Enhanced Stakeholder Engagement:

- Facilitates better communication and engagement with stakeholders by addressing concerns that are most relevant to them.

Nike has conducted a materiality assessment to prioritize its ESG issues. The assessment highlighted key areas such as sustainable materials, labor practices, and innovation in manufacturing processes based on their significant impact and importance to stakeholders.[xxiv]

Conducting a materiality assessment is crucial for organizations aiming to manage ESG issues effectively. It allows them to focus on the most impactful areas, align strategies with stakeholder expectations, and allocate resources efficiently, thus driving meaningful progress towards sustainability.

Conducting a thorough assessment is not merely an academic exercise; it's the compass guiding the journey toward sustainability. It provides a factual basis for formulating strategies, setting goals, and prioritizing initiatives that can drive meaningful change within the organization.

Liberty Mutual engages with various stakeholders, including employees, consumers, investors, brokers, agents, and non-governmental organizations (NGOs). In 2020, they conducted their first materiality assessment exercise to understand stakeholder priorities related to ESG topics. The assessment provided valuable insights that inform their ESG journey and communications with stakeholders.

Prior to engaging stakeholders, Liberty Mutual reviewed ESG reporting frameworks, regulatory requirements, and industry peer assessments. They defined a list of 14 of the most material ESG topics, covering both environmental and social aspects. These topics included diversity, equity, inclusion, human capital management, employee experience, health and wellness, climate change, environmental management, and more.[xxv]

Bank of America engages with diverse stakeholders, including its National Community Advisory Council (NCAC), to tailor its ESG disclosures. They utilize industry frameworks and ratings such as GRI, SASB, TCFD, CDP, and others to guide their sustainability reporting. Their internal Responsible Growth Committee oversees ESG priorities and strategy. The Bank's ESG materiality matrix ranks priorities based on their relative importance.[xxvi]

Conclusion

It is very important to assess the current state of a company's positive and negative impact on its surroundings as a foundational step towards sustainability. A clearer picture emerges through a systematic analysis of a company's competitive landscape, sustainability trends, regulatory environment, business opportunities, and future outlook. Understanding the strengths, weaknesses, opportunities, and threats from an ESG perspective allows for strategic decision-making and a proactive approach towards sustainability-driven growth.

The "As-is" analysis of a company's ESG impact delves deeper into its environmental, social, and governance practices, highlighting areas of excellence and opportunities for improvement. It provides a basis for goal setting, enhances transparency and reporting, and informs strategic decision-making. Additionally, data gathering related to ESG and conducting a materiality assessment helps identify and prioritize ESG issues that are most relevant and impactful.

Through these processes, organizations can navigate the complex landscape of sustainability, cultivate a culture of accountability, and drive continuous improvement. As organizations set up on this journey, it's not just about meeting regulatory requirements or ticking boxes; it's about holding sustainability as a core value, integrating it into the organizational culture and DNA, and ultimately contributing to a more sustainable and prosperous future for all.

Chapter 3
Set Clear Goals

"Goals may give focus, but dreams give power. Make sure yours are both clear and compelling." – John C. Maxwell.

In the pursuit of sustainability, setting clear and well-defined goals is the compass that guides an organization towards meaningful change. These goals act as a roadmap, aligning efforts with the company's values long-term vision and addressing the most material environmental and social impacts.

Here's why setting clear and well-defined goals is the pivotal compass steering you towards meaningful change:

1. The Importance of Goals & Targets:

Imagine setting sail without a destination. Goals provide that destination. They give purpose and direction to your sustainability initiatives. Without them, efforts may scatter, and the impact may lack focus. Goals act as the driving force, steering the collective energy of your organization towards a common vision of a sustainable future.

2. Supporting Ambitions Through Targets:

Ambitions are the overarching dreams; targets are the actionable steps that bring those dreams to life. Targets provide a practical roadmap, breaking down ambitious long-term visions into manageable, achievable milestones. They offer clarity on what needs to be done, when, and how, transforming lofty aspirations into actionable plans.

3. The Blend Of Quantitative & Qualitative Goals:

While quantitative goals offer clear metrics, don't underestimate the power of qualitative goals. Some impacts are not easily quantifiable but are equally crucial. Qualitative goals, such as fostering a culture of inclusivity or enhancing community engagement, complement quantitative metrics, providing a holistic view of your organization's impact.

4. Measurable Goals:

Measurability is the heartbeat of effective goal-setting. If you can't measure it, you can't manage it. Each goal should be accompanied by tangible metrics, enabling you to track progress and assess its impact. Whether it's reducing carbon emissions, increasing employee satisfaction, or improving supply chain sustainability, measurable goals provide accountability and a basis for informed decision-making.

5. Short-Term, Medium-Term, & Long-Term Goals:

Goals are not a one-size-fits-all endeavor. Breaking them into short-term, medium-term, and long-term categories aligns them with the pace of your organization's journey. Short-term goals, achievable within 12 months, offer quick wins and build momentum. Medium-term goals, spanning 2–3 years, delve deeper into systemic changes. Long-term goals, set for 5 years or more, embody your organization's enduring commitment to sustainability.

How This Benefits You:

- **Focus & Prioritization:** Goals provide a focal point, helping you prioritize efforts based on their impact and relevance.
- **Alignment With Stakeholder Expectations:** Clear goals resonate with stakeholders, showcasing your commitment to transparency and accountability.
- **Continuous Improvement:** Regularly evaluating and adjusting goals enables continuous improvement, ensuring your sustainability strategy stays agile and effective.
- **Demonstrating Progress:** Achieving goals, whether quantitative or qualitative, serves as a tangible testament to your organization's dedication to sustainability.

In conclusion, goals are not just checkpoints on a journey; they are the journey itself. The SMART framework—Specific, Measurable, Achievable, Relevant, and Time-Bound—provides a structured approach to goal-setting that ensures effectiveness and accountability.

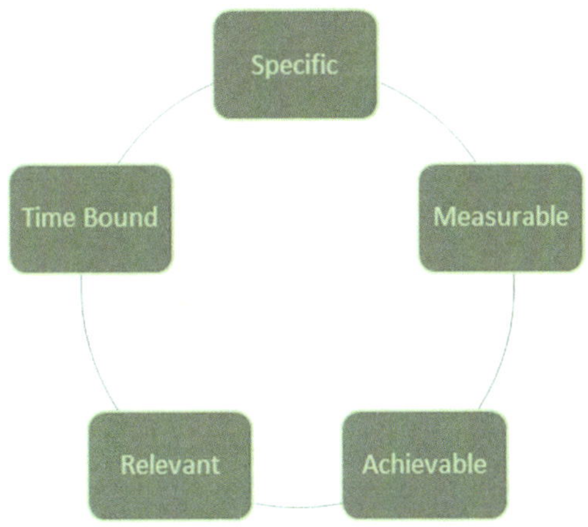

Specific:

Environmental Focus: Clearly define the environmental aspects that the goals aim to address. Whether it's reducing carbon emissions, minimizing water consumption, or optimizing resource usage, specificity ensures a targeted approach.

Coca-Cola has set a goal to replenish 100% of the water it uses in its beverages and production processes by 2030. They focus on sustainable water sourcing, watershed protection, and water recycling efforts to achieve this goal.[xxvii]

Social Objectives: Outline specific social impact objectives, such as improving employee well-being, enhancing community engagement, or fostering diversity and inclusion. Specificity ensures a clear understanding of the desired outcomes.

Google focuses on employee well-being through initiatives like providing comprehensive healthcare benefits, promoting work-life balance, offering wellness programs, and fostering a supportive and inclusive work culture.[xxviii]

Measurable:

Quantifiable Metrics: Establish measurable metrics to gauge progress. Whether it's reducing energy consumption by a certain percentage, achieving zero waste targets, or increasing employee satisfaction scores, having measurable indicators allows for tracking and evaluation.

Data-Driven Evaluation: Use data to assess the organization's current state and measure the impact of sustainability initiatives. This provides a foundation for setting realistic and achievable targets.

With the goal of achieving zero waste in landfills at all final assembly sites by 2030, Apple has committed to sending no waste to landfills at its final assembly sites by 2030. They measure waste generation, implement recycling and reuse programs, and work with suppliers to minimize waste throughout their production processes.[xxix]

Achievable:

Realistic Targets: Goals should be ambitious yet attainable. Consider the organization's capabilities, available resources, and industry benchmarks when determining what is achievable.

Stepwise Approach: Break down larger sustainability goals into smaller, manageable milestones. This facilitates a step-by-step approach, making the overall journey more achievable and less overwhelming.

Google has committed to investing $1 billion in grants, training programs, and partnerships to support economic opportunity and job creation in underserved communities. They are focusing on initiatives that promote digital skills development, entrepreneurship, and small business growth.[xxx]

Relevant:

Materiality Alignment: Ensure that sustainability goals align with the most material environmental and social impacts. This alignment enhances relevance, focusing efforts on issues that truly matter to stakeholders and the organization's long-term viability.

Strategic Alignment: Connect sustainability goals with the company's overall business strategy. The relevance of these goals should contribute to the organization's mission, vision, and values.

Unilever's goal of halving the use of virgin plastic in packaging by 2025 aligns with the material environmental impact of plastic waste. This goal is relevant as it addresses a pressing global issue and resonates with stakeholders concerned about environmental conservation.[xxxi]

Time-Bound:

Set Timelines: Define clear timelines for achieving sustainability goals. Whether it's short-term targets or long-term aspirations, having deadlines fosters a sense of urgency and accountability.

Regular Evaluation: Establish periodic reviews to assess progress and make adjustments as needed. Timely evaluations ensure that the organization stays on course and adapts to changing circumstances.

IKEA's long-term aspiration to become fully circular and climate-positive by 2030 sets a definitive deadline for transitioning its operations to sustainable and regenerative practices. Regular evaluations ensure progress towards this ambitious goal is tracked and adjustments are made as necessary.[xxxii]

Unilever's Sustainable Living Plan is a great example of how companies can set clear and SMART sustainability goals. The plan was launched in 2010 and has three main goals: to help more than a billion people improve their health and well-being, to halve the environmental impact of the making and use of their products, and to enhance the livelihoods of thousands of people in their supply chain.[xxxiii] The plan has hundreds of targets and commitments, all with specific, measurable, achievable, relevant, and time-bound criteria.

Unilever has made significant progress towards achieving these goals. For example, they have reached 1.3 billion people through their health and hygiene programs.[xxxiv] They have also reduced their greenhouse gas emissions by 43% per ton of production since 2008.[xxxv]

Setting clear and SMART sustainability goals not only charts the course for an organization's sustainability journey but also provides a framework for continuous improvement and impactful change.

Goals & Ambitions Within ESG

Setting tangible goals and ambitions within the Environmental (E), Social (S), and Governance (G) dimensions is a critical step in a company's sustainability journey. These goals serve as benchmarks, guiding the organization towards positive impact and responsible business practices.

As you set sail on your sustainability journey, envision your goals as beacons that illuminate not only the immediate path but also the distant horizons. The dual approach of setting short-term goals and long-term ambitions within the Environmental (E), Social (S), and Governance (G) dimensions is the compass that ensures your organization's sustainability voyage is both purposeful and enduring.

Short-Term Goals – Immediate Wins With Lasting Impact:

Short-term goals, typically achievable within a span of 12 months, serve as your immediate milestones. They are the quick wins that demonstrate your organization's commitment to positive change. These goals are crucial for building momentum, fostering a sense of accomplishment, and showcasing the tangible impacts of your sustainability initiatives.

- Achieving short-term goals provides a morale boost and builds momentum within your organization. It reinforces the belief that positive change is not only possible but already underway.
- Short-term goals are effective in addressing urgent issues that demand immediate attention. Whether it's reducing energy consumption, optimizing waste management, or enhancing employee well-being, these goals tackle pressing challenges head-on.
- Successfully meeting short-term goals enhances your organization's credibility. It demonstrates to stakeholders, including employees, customers, and investors, that your sustainability commitment is not just rhetoric but backed by tangible actions.

Long-Term Ambitions – Guiding the Journey Towards Enduring Impact:

Long-term ambitions, set for a horizon of 5 years or more, define the overarching vision for your organization's sustainability journey. They are the North Star, guiding your strategic decisions and shaping your organizational culture. While short-term goals provide immediate wins, long-term ambitions ensure that your organization's impact is not only immediate but also enduring.

- Long-term ambitions set the strategic direction for your sustainability journey. They encapsulate the larger vision, guiding your organization through systemic changes and transformative initiatives that go beyond incremental improvements.
- By focusing on long-term ambitions, your organization institutionalizes sustainability into its core values and operations. It becomes an integral part of decision-making processes, product development, and stakeholder engagement, fostering a culture of responsibility.
- While short-term goals address immediate challenges, long-term ambitions enable you to tackle the root causes of sustainability issues. This comprehensive approach ensures that your organization is not merely reacting to problems but proactively shaping a more sustainable future.

Balancing the Time Horizons:

The synergy between short-term goals and long-term ambitions is essential. Short-term wins provide the fuel for the journey, while long-term ambitions provide the roadmap. This balanced approach ensures that your organization reaps immediate benefits while continuously progressing towards a more sustainable and responsible future.

Essential Insight:

Balancing Short-Term Goals with Long-Term Ambitions is a Unified Strategy. Immediate successes offer motivation and validation, while enduring ambitions provide a strategic roadmap for sustained impact. This combined approach propels organizations towards a purposeful and forward-thinking sustainability journey.

Ambitions:

Setting medium- to long-term ambitions is crucial because it establishes the strategic direction and vision for an organization. These ambitions provide a clear roadmap for where the organization wants to be in the future, guiding decision-making, resource allocation, and goal-setting processes.

Long-term ambitions ensure that sustainability efforts are aligned with the organization's overall strategic goals and mission. This alignment prevents disjointed initiatives and ensures that sustainability becomes an integral part of the organization's core strategy. They inspire employees, stakeholders, and investors by painting a compelling vision of the organization's future. They create a sense of purpose and direction, motivating individuals to contribute towards achieving these ambitious goals.

Medium- to long-term ambitions help in prioritizing investments and allocating resources effectively. Organizations can identify key areas where investments are needed to drive sustainability initiatives forward, ensuring optimal use of resources.

Clear, long-term ambitions attract stakeholders who are aligned with the organization's vision and values. This engagement can lead to partnerships, collaborations, and support from stakeholders who share a common interest in achieving sustainability goals.

While long-term ambitions provide a strategic direction, they also allow for adaptability and flexibility in implementation. Organizations can adjust strategies and tactics along the way to respond to changing external factors or emerging opportunities without losing sight of the overarching vision.

Setting a 3-5 year ambition for a company's sustainability journey involves outlining broader, aspirational goals that align with the organization's values, long-term vision, and the overarching impact it seeks to achieve. These ambitions provide a sense of direction, guiding the company's efforts towards significant and transformative outcomes within the Environmental (E), Social (S), and Governance (G) dimensions.

1. Long-Term Ambitions Embrace the Visionary Horizon

Long-term ambitions encapsulate the organization's visionary aspirations for sustainability. These are overarching goals that often extend beyond the traditional planning horizon, reaching into the next decade or more.

For example, achieving carbon neutrality by 2050, transitioning to 100% renewable energy, or becoming a zero-waste enterprise.

Long-term ambitions necessitate transformative change. They often involve reimagining fundamental aspects of the organization's operations, products, and relationships to align with a sustainable future.

For example, redesigning products for full recyclability, implementing regenerative agriculture practices or creating a closed-loop supply chain.

These ambitions extend beyond organizational boundaries, aiming to contribute positively to global sustainability challenges. Long-term ambitions may involve influencing industry standards, advocating for policy changes, or pioneering innovative solutions.

For example, leading industry-wide initiatives for sustainable sourcing, partnering with global organizations on climate action, or advocating for biodiversity conservation.

Environmental Ambitions (E):
Renewable Energy Transition

- o **3-5 Year Ambition:** Transition to a majority reliance on renewable energy sources for operations, significantly reducing the carbon footprint.
- o **Strategic Actions:** Invest in solar and wind energy projects, explore power purchase agreements (PPAs) with renewable energy providers, and implement energy-efficient technologies.

Circular Economy Integration

- o **3-5 Year Ambition:** Embed circular economy principles into product design and supply chain operations, minimizing waste and promoting sustainable resource use.
- o **Strategic Actions:** Establish closed-loop systems, explore product redesigns for recyclability, and collaborate with suppliers for responsible sourcing and disposal.

Biodiversity Conservation

- o **3-5 Year Ambition:** Implement biodiversity conservation initiatives, ensuring that business activities have a positive or neutral impact on ecosystems.
- o **Strategic Actions:** Conduct biodiversity assessments, implement habitat restoration projects, and engage in partnerships with environmental organizations.

Social Ambitions (S):
Employee Well-Being Excellence

- o **3-5 Year Ambition:** Attain recognition as an employer of choice with industry-leading employee satisfaction and well-being programs.
- o **Strategic Actions:** Enhance wellness initiatives, provide professional development opportunities, and foster a supportive and inclusive workplace culture.

Community Empowerment

- o **3-5 Year Ambition:** Strengthen ties with local communities, contributing to their socio-economic development through impactful projects.
- o **Strategic Actions:** Collaborate on community development programs, engage in transparent communication, and establish partnerships with local organizations.

Diversity & Inclusion Leadership

- o **3-5 Year Ambition:** Become a benchmark for diversity and inclusion, achieving a diverse and inclusive workforce at all levels.
- o **Strategic Actions:** Implement inclusive hiring practices, offer diversity training, and establish mentorship programs.

Governance Ambitions (G):
Ethical Leadership Recognition

- o **3-5 Year Ambition:** Achieve recognition for ethical business practices, transparency, and accountability in governance.
- o **Strategic Actions:** Strengthen anti-corruption measures, enhance reporting transparency, and seek industry-specific certifications for ethical excellence.

Board Diversity and Expertise

- o **3-5 Year Ambition:** Attain a diverse and skilled board of directors, reflective of the organization's commitment to inclusive governance.
- o **Strategic Actions:** Implement targeted recruitment for diverse board members, provide training on ESG matters, and establish governance committees focused on sustainability.

Shareholder Trust & Engagement

- o **3-5 Year Ambition:** Cultivate a shareholder base that actively supports the company's sustainability journey through transparent communication and active engagement.
- o **Strategic Actions:** Develop regular communication channels with shareholders, provide updates on sustainability progress, and seek input on key ESG initiatives.

Unilever's Sustainable Living Plan is a comprehensive plan that sets out hundreds of targets and commitments to help more than a billion people improve their health and well-being, halve the environmental impact of the making and use of their products, and enhance the livelihoods of thousands of people in their supply chain.[xx]

Unilever has set a 3-5 year ambition to achieve its Sustainable Living Plan targets, which include sourcing 100% of agricultural raw materials sustainably and improving the health and well-being of one billion people through its products.[xx]

To achieve these targets, Unilever has implemented various initiatives. For instance, they have launched the Sustainable Agriculture Code, which provides guidelines for sustainable farming practices. They have also introduced the Unilever Sustainable Living Plan Progress Report, which tracks their progress towards their sustainability goals.[xxi]

Setting 3-5 year ambitions provides a roadmap for transformative change, guiding the company towards becoming a more sustainable and responsible entity. These ambitions, when coupled with strategic actions and continuous evaluation, propel the organization towards a future where it not only meets but exceeds its ESG commitments.

2. Medium-Term Goals Bridge Vision & Action

Medium-term ambitions bridge the visionary with the actionable. They represent milestones and targets that the organization aims to achieve within a more immediate timeframe, typically spanning three to five years.

For example, reducing operational carbon emissions by 30% within the next five years, achieving a specific percentage increase in sustainable sourcing, or enhancing product eco-design.

Medium-term ambitions often involve the integration of sustainable practices into day-to-day operations. They may also encompass innovation initiatives aimed at developing and adopting new technologies or practices.

For example, implementing energy-efficient technologies in manufacturing processes, launching sustainable product lines, or enhancing employee engagement in sustainability initiatives.

Collaboration with stakeholders becomes crucial during the medium term, as ambitions require alignment with the expectations and needs of customers, employees, suppliers, and communities.

For example, collaborating with suppliers to establish a responsible sourcing framework, engaging employees in carbon reduction initiatives, or partnering with local communities on social impact projects.

Goals and Their Importance in Supporting Ambitions

Goals serve as the stepping stones that bridge an organization's ambitions with actionable plans and tangible outcomes. They provide a structured framework for turning long-term visions into manageable tasks and measurable achievements. Setting clear and specific goals is essential to supporting the ambitions set by organizations, particularly in the realm of sustainability. These goals serve several crucial purposes that are instrumental in driving progress and realizing strategic ambitions.

Goals bring clarity and focus to the broader ambitions of an organization. They define what needs to be accomplished, by when, and with what resources, offering a roadmap that guides decision-making and operational efforts.

Well-defined goals are measurable, allowing organizations to track progress and assess performance objectively. This measurability fosters accountability as stakeholders can evaluate whether targets are being met and take corrective actions if needed.

Goals provide a sense of purpose and motivation for employees, stakeholders, and partners. Clear objectives create a shared understanding of priorities and inspire collective efforts towards achieving common sustainability goals.

Goals help in prioritizing resource allocation by identifying key areas that require investment, expertise, and attention. This strategic allocation ensures that resources are utilized efficiently to drive sustainable outcomes.

While goals provide a structured framework, they also allow for adaptability and iteration. Organizations can adjust goals based on changing circumstances, emerging opportunities, or new insights, ensuring agility in pursuing sustainability ambitions.

Environmental Goals (E):

Reducing Carbon Footprint

- o **Tangible Goal:** Achieve a specific percentage reduction in carbon emissions within the next 3–5 years, in alignment with the company's as-is analysis and market trends.
- o **Ambition:** Strive for carbon neutrality by investing in renewable energy sources, energy-efficient technologies, and sustainable practices across operations.

Resource Efficiency

- o **Tangible Goal:** Decrease water consumption and optimize resource usage by a measurable amount, promoting a circular economy.
- o **Ambition:** Establish closed-loop systems for key resources, minimizing waste and maximizing resource efficiency throughout the supply chain.

Waste Reduction

- o **Tangible Goal:** Implement strategies to reduce waste generation by a certain percentage, incorporating recycling and waste diversion initiatives.
- o **Ambition:** Work towards achieving zero waste in landfills, actively promoting a culture of waste reduction and responsible disposal.

Social Goals (S):

Employee Well-being

- o **Tangible Goal:** Enhance employee well-being by implementing wellness programs, flexible work arrangements, and promoting a positive workplace culture.
- o **Ambition:** Strive to be recognized as an employer of choice with industry-leading employee satisfaction scores.

Community Engagement

- o **Tangible Goal:** Strengthen ties with local communities through impactful philanthropy, community development projects, and transparent communication.
- o **Ambition:** Establish the company as a responsible corporate citizen deeply integrated into the fabric of the communities it serves.

Diversity & Inclusion

- o **Tangible Goal:** Increase diversity representation at all organizational levels through targeted hiring and inclusive policies.
- o **Ambition:** Achieve industry-leading diversity metrics and foster an inclusive workplace culture that values and celebrates differences.

Governance Goals (G):

Ethical Business Practices

- o **Tangible Goal:** Strengthen ethical business practices by enhancing transparency, accountability, and adherence to anti-corruption measures.
- o **Ambition:** Achieve recognition for ethical excellence through industry awards and third-party certifications.

Board Diversity

- o **Tangible Goal:** Increase diversity within the board of directors, ensuring representation across gender, ethnicity, and skill sets.
- o **Ambition:** Strive to be a benchmark for board diversity, setting industry standards for inclusive governance.

Shareholder Engagement

- o **Tangible Goal:** Establish mechanisms for regular and meaningful engagement with shareholders, seeking input on sustainability practices.
- o **Ambition:** Cultivate a shareholder base that actively supports and champions the company's commitment to ESG principles.

Microsoft's Environmental Sustainability Goals[xxxvi]: Microsoft has set an ambitious goal to become carbon-negative by 2030 and remove all historical carbon emissions by 2050. They aim to achieve this by adopting a three-pronged approach: being carbon-negative, water-positive, and zero-waste by 2030.

To achieve its carbon-negative goal, Microsoft plans to remove more carbon from the atmosphere than it emits. They also plan to invest in carbon removal and recycling to reduce carbon emissions across their supply chain.

In addition to its carbon-negative goal, Microsoft aims to be water-positive by 2030, meaning it will replenish more water than it uses. They also aim to be zero-waste by 2030, which means they will eliminate waste across their direct waste footprint.

Microsoft is committed to protecting and preserving ecosystems. They plan to protect more land than they use by 2025 and build a planetary computer.

Microsoft is also using AI to accelerate sustainability by helping organizations worldwide meet their own sustainability objectives that are aligned with global targets.

Setting tangible goals within ESG dimensions not only demonstrates a commitment to sustainability but also provides a clear roadmap for action. These goals, when pursued with ambition, propel the organization towards a future where environmental, social, and governance considerations are integral to its identity and success.

KPI's Navigate Immediate Impact

Short-term ambitions focus on immediate impact and tangible outcomes. They often involve initiatives that can be implemented swiftly, providing quick wins that build momentum for broader sustainability efforts.

For example, implementing energy-saving measures in office facilities, initiating waste reduction programs, or conducting employee training on sustainability practices.

Short-term ambitions may also target behavioral shifts within the organization, fostering a culture of sustainability among employees and stakeholders.

For example, launching internal campaigns to reduce single-use plastics, encouraging public transportation or remote work options, or initiating a workplace recycling program.

Setting Key Performance Indicators (KPIs) for short-term ambitions ensures that progress can be measured and adjustments can be made swiftly. These KPIs provide a feedback loop for ongoing improvement.

For example, tracking monthly energy consumption reductions, measuring waste diversion rates, or monitoring employee participation in sustainability initiatives.

Yearly Targets & KPIs

Setting yearly targets and Key Performance Indicators (KPIs) is a crucial step in operationalizing sustainability ambitions. These targets provide a more granular roadmap for achieving broader ambitions and enable organizations to measure progress, assess performance, and make data-driven decisions within the Environmental (E), Social (S), and Governance (G) dimensions.

As you move from commitment to action, these granular metrics become your compass, guiding your organization's every move towards a more sustainable and responsible future.

Yearly KPIs Bridge the Gap Between Commitment & Action.

While overarching commitments set the tone, yearly KPIs bridge the gap between aspiration and implementation. They provide a tangible, measurable framework that transforms high-level goals into operational realities. These yearly KPIs are the heartbeat of your sustainability strategy, marking each year's progress and illuminating the path forward.

- **Operational Focus:** Yearly KPIs shift the focus from broad, long-term goals to specific, actionable steps. They break down the journey into manageable, achievable milestones, ensuring that each year brings measurable progress.
- **Adaptability & Flexibility:** In the dynamic landscape of sustainability, yearly KPIs offer adaptability. They allow your organization to respond to changing circumstances, emerging trends, and evolving stakeholder expectations, ensuring that your sustainability efforts remain relevant and effective.
- **Demonstrating Continuous Improvement:** Yearly KPIs are not just about meeting targets; they showcase your organization's commitment to continuous improvement. Regularly assessing and recalibrating these metrics fosters a culture of agility and responsiveness to emerging sustainability challenges.

Adding KPIs to the Company or Management Balance Scorecard Is Important.

Integrating sustainability KPIs into the company or management balance scorecard elevates their significance. It places sustainability metrics on par with financial and operational indicators, reinforcing the message that sustainability is integral to overall business success.

- **Holistic Performance Measurement:** A balanced scorecard that includes sustainability KPIs offers a holistic view of your organization's performance. It communicates to stakeholders that sustainability is not a standalone initiative but an integrated aspect of your overall business strategy.

- **Strategic Alignment:** By aligning sustainability KPIs with other key performance indicators, you communicate the interconnectedness of sustainability and organizational success. This integration ensures that sustainability is not treated as a siloed effort but is woven into the fabric of strategic decision-making.

- **Board & Leadership Oversight:** Including sustainability KPIs in the balanced scorecard brings these metrics under the purview of boards and leadership teams. This oversight ensures that sustainability remains a boardroom discussion, receiving the attention and resources necessary for success.

Linking Remuneration to the KPIs

Elevating the importance of sustainability KPIs can be further reinforced by linking remuneration to their achievement. This strategic alignment ensures that sustainability performance is not only a strategic objective but a fundamental component of individual and team success.

The importance of sustainability KPIs in executive remuneration is underscored by recent trends in corporate practices. According to a study by WTW, a US advisory and broking company, there has been a notable increase in the incorporation of ESG metrics into executive pay plans among UK companies, reaching 89% from 81% previously. This inclusion spans both short-term and long-term payment schemes, with 85% of companies aligning short-term pay packages with at least one ESG measure, up from 79%. Similarly, the use of ESG measures in long-term pay plans has grown from 24% to 37% over the past year.[xxxvii]

These statistics highlight a growing recognition among companies of the need to integrate sustainability goals into their compensation structures. Linking remuneration to the achievement of sustainability KPIs not only elevates the strategic importance of these objectives but also incentivizes individual and team success based on sustainability performance. As Sarah Reay, Climate Change Executive at ICAEW, notes, incorporating social and environmental measures into executive remuneration can catalyze action on emissions reduction and social inequalities and overall improve societal and environmental well-being. With Europe leading this trend, particularly countries like Germany and France, it is expected that ESG-linked incentive plans will become increasingly commonplace globally, driven by investor and regulatory pressures to prioritize sustainability in corporate decision-making.

One notable example is **Unilever**, which has integrated sustainability metrics into its executive pay structure. Unilever's Compensation Committee, in its 2022 Directors' Remuneration Report (DRR)[xxxviii], highlights the strategic linkage between business performance and remuneration outcomes. Despite challenging macroeconomic conditions, Unilever achieved strong growth, with metrics such as underlying sales growth (USG) at 9.0% and underlying operating margin (UOM) slightly ahead of target at 16.1%. These achievements were recognized through the determination of annual bonuses for the CEO and CFO based on formulaic outcomes aligned with sustainability goals.

For instance, CEO Alan Jope and CFO Graeme Pitkethly received bonuses of 200% and 160% of fixed pay, respectively, against their target opportunities. The Committee attributed these outcomes to Unilever's emphasis on sustainability, as evidenced by upweighting USG to 50% within the 2022 annual bonus performance measures. This decision reflects Unilever's commitment to driving sustainable growth by balancing financial performance with environmental and social considerations.

Moreover, Unilever's approach extends beyond executive remuneration to encompass broader workforce incentives. The Management Co-Investment Plan (MCIP) for eligible employees saw a discretionary adjustment of +10% to the formulaic outcome, reflecting the impact of external factors like COVID and input cost inflation. This demonstrates Unilever's efforts to align remuneration practices with wider stakeholder considerations, ensuring fairness and transparency.

- **Incentivizing Sustainability Leadership:** Linking remuneration to sustainability KPIs creates a powerful incentive for leadership to champion sustainability initiatives. It signals that achieving sustainability goals is not just a collective responsibility but a factor in individual and team success.

- **Aligning Personal Goals With Organizational Values:** When remuneration is tied to sustainability KPIs, it encourages individuals at all levels to align their personal goals with the organization's values. This alignment fosters a sense of shared purpose, motivating employees to contribute to sustainability objectives actively.

- **Demonstrating Commitment to Stakeholders:** By linking remuneration to sustainability KPIs, organizations convey a commitment to stakeholders that goes beyond rhetoric. It demonstrates a tangible commitment to embedding sustainability principles in the very fabric of the organization.

Environmental Yearly Targets & KPIs (E):

- **Yearly Carbon Emission Reduction Target:**

  - **Target:** Achieve a specific percentage reduction in carbon emissions compared to the previous year.
  - **KPIs:** Measure emissions across scopes (Scope 1, Scope 2, and Scope 3), track energy consumption, and assess the impact of renewable energy initiatives.

- **Energy Efficiency Improvement:**

  - **Target:** Reduce energy consumption by a certain percentage through efficiency measures.
  - **KPIs:** Monitor energy usage in facilities, track the implementation of energy-efficient technologies, and assess the impact on overall energy efficiency.

- **Waste Reduction Target:**

  - **Target:** Decrease waste generation by implementing recycling and waste reduction strategies. The target of percentage of waste recycled.

- o **KPIs:** Monitor waste generation rates, track recycling percentages, and assess the implementation of circular economy principles. Percentage of waste

- Other Environmental KPI's to be considered based on industry, etc.:

  - o Volume of water consumed relative to each unit of production/service
  - o Y/Y development in plastic used in operations or packaging
  - o Number of initiatives/their effectiveness in preserving or enhancing biodiversity
  - o Assessment of major suppliers based on their environmental impact
  - o Investment/financial commitment towards environmental management activities
  - o Climate risk evaluations to identify and manage business risk from climate change

Social Yearly Targets & KPIs (S):

- **Employee Well-Being Enhancement Target:**

  - o **Target:** Improve employee satisfaction scores and well-being metrics.
  - o **KPIs:** Conduct regular employee surveys, track participation in wellness programs, and assess turnover rates and absenteeism.

- **Community Impact Target:**

  - o **Target:** Implement specific community projects or initiatives to impact local communities positively.
  - o **KPIs:** Monitor the progress and outcomes of community engagement projects, assess community feedback, and measure the social impact.

- **Diversity & Inclusion Improvement Target:**

  o **Target:** Increase diversity representation at various organizational levels.
  o **KPIs:** Track diversity metrics in recruitment and promotions, assess the success of diversity training programs and measure the effectiveness of inclusion initiatives.

- Other Social KPI's to be considered based on industry etc.:

  o Average training hours per employee per year
  o Percentage of total cost invested in innovation
  o Percentage of employees leaving the company
  o Percentage of employees undergoing annual performance evaluations
  o Count of health and safety issues reported
  o Average score from the customer feedback survey
  o Number of cases where customer data privacy was compromised
  o Time taken to address and close customer complaints
  o Share of net profit allocated to CSR initiatives
  o Number of cases reported of discrimination, harassment, modern slavery
  o Number of partnerships with NGO's or community services

Governance Yearly Targets & KPIs (G):

- **Ethical Business Practices Target:**

  o **Target:** Strengthen ethical business practices and transparency.
  o **KPIs:** Monitor compliance with anti-corruption measures, assess transparency in reporting, and track adherence to ethical standards.

- **Board Diversity Improvement Target:**

  o **Target:** Enhance board diversity in terms of gender, ethnicity, and expertise.

- o **KPIs:** Track changes in board composition, assess the success of targeted recruitment efforts and monitor diversity training for board members.

- **Shareholder Engagement Target:**

  - o **Target:** Strengthen communication and engagement with shareholders on sustainability matters.
  - o **KPIs:** Monitor shareholder meetings and feedback mechanisms, assess the level of shareholder satisfaction, and track changes in shareholder sentiment.
- Other Governance KPI's to be considered based on industry, etc.:

  - o Implementation of a formal risk management framework
  - o Number of cybersecurity threats or breaches reported
  - o Proportion of reporting guidelines complied with
  - o Assessment of the effectiveness of internal controls
  - o Share of audit recommendations implemented

Nike's FY25 Targets: Nike has set ambitious sustainability targets for 2025, including achieving a 30% reduction in carbon emissions across its global supply chain. KPIs include tracking emission reductions, evaluating sustainable material usage, and assessing progress in water stewardship.[xxxix]

Yearly targets and KPIs are instrumental in translating sustainability ambitions into actionable steps. They provide a structured approach to measuring progress, identifying areas for improvement, and ensuring that the organization remains on course to achieve its overarching ESG goals.

Steps To Remember When Setting Clear and Effective Goals

Setting clear and effective sustainability goals is a cornerstone of a purpose-driven and responsible organization. The success of these goals hinges on their clarity, alignment with the organization's vision and values, specificity, realism, timeframes, relevance, ownership, resource allocation, risk assessment, and a robust monitoring process.

1. Clear Communication:

Clarity is paramount. Each sustainability goal should be expressed in simple and understandable terms, ensuring that every member of the organization, from the boardroom to the front lines, comprehends the objectives. This clarity fosters a shared understanding of the company's sustainability direction.

2. Alignment With Vision, Mission, & Core Values:

Sustainability goals must seamlessly align with the organization's broader vision, mission, and core values. This alignment not only reinforces the company's commitment to responsible business practices but also ensures that sustainability becomes an integral part of its identity.

3. Specificity for Measurability:

Specificity transforms aspirations into measurable targets. Each goal should be clearly defined, allowing for precise measurement. This specificity enables the organization to track progress effectively, providing insights into what is working and where adjustments are needed.

4. Realistic Ambitions:

Goals should strike a balance between ambition and realism. While being challenging and ambitious, they must remain attainable. Realistic goals inspire continuous improvement without setting the organization up for unachievable targets, ensuring sustained commitment and motivation.

5. Timeframes for Urgency:

Assigning clear timeframes to each goal creates a sense of urgency. A deadline instils focus and discipline, prompting proactive efforts to achieve the desired outcomes within a defined period. Timeframes also facilitate periodic evaluations and course corrections.

6. Relevance to Business Environment:

Sustainability goals should be rooted in the business environment, addressing key opportunities and challenges. They should resonate with the company's industry, stakeholder expectations, and societal needs, ensuring that efforts align with broader environmental, social, and governance imperatives.

7. Ownership & Accountability:

Assigning clear ownership of each goal to specific individuals or teams is crucial. This ensures accountability, making it evident who is responsible for driving progress. Ownership fosters a sense of responsibility and empowers individuals to take meaningful action.

8. Resource Allocation:

Understand the resources required for goal attainment. Whether financial, human, technological, or other resources, a clear understanding ensures that the necessary support is in place, preventing bottlenecks and promoting a smooth execution process.

9. Risk Assessment:

Identify potential risks and challenges that may impede goal achievement. A comprehensive risk assessment allows the organization to proactively address obstacles, develop mitigation strategies, and navigate uncertainties in the pursuit of sustainability.

10. Monitoring & Review Process:

Establish a robust process for monitoring and reviewing progress towards the goals. Regular assessments provide feedback loops, allowing the organization to adapt strategies, address emerging challenges, and celebrate successes along the sustainability journey.

In conclusion, the effectiveness of sustainability goals lies not only in their formulation but also in their commitment to their execution. Each goal becomes a beacon guiding the organization towards a future where responsible business practices are not just ideals but tangible, measurable outcomes.

Future Outlook Based on As-Is & Market Analysis

Envisioning the future trajectory of a company based on the as-is analysis and market analysis is a strategic exercise that involves leveraging insights gained from these assessments to chart a course for sustainable growth and positive impact.

1. **Leveraging Strengths & Addressing Weaknesses:**

- **Building on Environmental Stewardship:** If the as-is analysis identifies the company's strengths in environmental practices, such as efficient resource usage or low emissions, the next 3-5 years could involve further advancements in sustainability, potentially becoming an industry leader in eco-friendly practices.
- **Mitigating Social Impact Risks:** If social impact is identified as a weakness, future strategies might focus on strengthening community engagement, improving labor practices, and fostering diversity and inclusion to enhance the company's social responsibility.

2. **Market Expansion & Positioning:**

- **Capitalizing on Sustainability Trends:** If market analysis reveals growing consumer preferences for sustainable products, the company could expand its offerings in this direction, capturing a larger market share and appealing to environmentally conscious consumers.
- **Navigating Regulatory Shifts:** Anticipating regulatory trends highlighted in the market analysis allows the company to proactively adapt to changing compliance requirements, ensuring it stays ahead of regulatory shifts.

3. **Strategic Goal Alignment:**

- **Setting Ambitious Sustainability Goals:** The future vision could involve setting ambitious sustainability goals aligned with material ESG factors. For example, if carbon emissions were identified as a significant concern, the company might set targets for carbon neutrality or reduction.

- **Strategic Materiality Integration:** Integrating materiality findings into strategic decision-making ensures that the company's focus areas align with stakeholder expectations, industry standards, and societal needs.

4. Operational Integration & Efficiency:

- **Embedding Sustainability in Operations:** The next 3–5 years could see the integration of sustainability into core business operations. This includes sustainable sourcing, eco-efficient manufacturing processes, and supply chain optimization to reduce environmental impact.
- **Continuous Improvement Culture:** Based on the as-is analysis, if operational inefficiencies were identified, the future outlook might involve fostering a culture of continuous improvement where employees are empowered to contribute to sustainability initiatives.

5. Stakeholder Trust & Collaboration:

- **Transparent Communication:** Building on the importance of transparency highlighted in the as-is analysis, the company could prioritize open communication about its sustainability journey. This fosters trust among stakeholders, including customers, investors, and communities.
- **Collaboration for Impact:** Recognizing opportunities for collaboration mentioned in the market analysis, the company could engage in partnerships with other organizations, NGOs, or government bodies to amplify the impact of its sustainability initiatives.

IKEA exemplifies the integration of as-is analysis and market analysis into its future strategy. After identifying the environmental impact of its operations, the company set ambitious goals to become climate-positive by 2030, aligning with market trends and stakeholder expectations.[xl]

IKEA's sustainability strategy, called "People & Planet Positive," aims to help make people's homes more sustainable and provide a better life for people and their communities. The strategy has three major focus areas: **Healthy & Sustainable Living, Circular & Climate Positive, and Fair & Equal**.

The future trajectory of the company, informed by the as-is analysis and market analysis, is not merely a projection but a strategic plan grounded in a thorough understanding of current realities and anticipated market dynamics. This forward-looking approach positions the company as a proactive and responsible player in the evolving landscape of sustainability.

Conclusion

In conclusion, establishing clear and well-defined sustainability goals is crucial for guiding an organization's journey towards a more sustainable and responsible future. By setting SMART (Specific, Measurable, Achievable, Relevant, and Time-bound) goals, companies can align their environmental, social, and governance ambitions with their long-term vision and values.

The foundation of these goals should be informed by comprehensive assessments, such as the "As-is" analysis and market analysis, which provide a clear picture of the company's current state and the external landscape. Based on this understanding, organizations can define where they want to be in 3-5 years and set tangible objectives within each of the ESG pillars.

It's crucial to set both long-term ambitions and yearly targets with specific Key Performance Indicators (KPIs). These goals should be easily understood by all stakeholders, clearly linked to the organization's core values, and have assigned ownership to ensure accountability. They should also be realistic yet ambitious, with a clear sense of urgency driven by realistic timeframes.

Furthermore, goal setting should consider the necessary resources, potential risks, and challenges and establish a robust monitoring and review process. This ongoing monitoring and review allows for adjustments to be made based on progress, ensuring that the organization remains on track towards its sustainability objectives.

In essence, setting clear sustainability goals is not only a strategic exercise but a fundamental commitment to the organization's stakeholders, its employees, its customers, and the planet. It signifies a purpose-driven approach to business that aims for positive societal impact while driving growth and innovation.

Chapter 4
Stakeholder Engagement

"The challenge of sustainable development is the challenge of creating a world that is better for everyone, not just a select few. It is about creating a world that is sustainable for generations to come. This requires the harmonious collaboration of diverse voices, where stakeholders are not just heard but actively engaged in the composition of meaningful change."

Effective stakeholder engagement lies at the heart of a sustainable and responsible business strategy. By identifying and actively involving key internal and external stakeholders, an organization can gain valuable insights, build trust, and ensure that its sustainability initiatives are aligned with the expectations and needs of those it impacts.

Identifying & Engaging with Key Stakeholders

Identifying key stakeholders involves recognizing the interconnected relationships that shape and are shaped by the organization's actions. This comprehensive mapping lays the foundation for effective engagement strategies, ensuring that sustainability initiatives are not only impactful but also reflective of the diverse interests and expectations within and around the organization.

Engagement is the bridge that connects organizational aspirations with the expectations and needs of key stakeholders. By actively involving internal and external stakeholders, an organization not only gains valuable insights but also forges relationships that are pivotal for the success of sustainability initiatives.

1. Employees:

ESG goals are imperative not only for a company's reputation and investor relations but also significantly impact employee attraction, engagement, and retention. According to a recent IBM study, 70% of employees find sustainability programs make employers more appealing, and 80% want to help their company reach climate or ESG goals.[xli] This data underscores the critical link between ESG initiatives and employee satisfaction, highlighting why ESG is crucial for employees.

Employees today are increasingly prioritizing companies with strong ESG commitments. They are drawn to organizations that align with their values and offer opportunities to contribute meaningfully to sustainability efforts. For instance, two in five respondents in a global survey of Gen Z and Millennials reported rejecting job offers or assignments that did not align with their values.[xli] This data showcases the importance of integrating values into everyone's role and how business gets done.

Moreover, employees who connect with the purpose behind their work are less likely to experience burnout or feel overwhelmed. This connection to purpose and values is a significant driver of employee engagement and retention. Research also indicates that workers satisfied with their company's social impact are more likely to stay with their employer for more than five years.[xli] This demonstrates the tangible benefits of creating opportunities for employees to engage with ESG goals.

Additionally, upskilling employees to contribute effectively to ESG initiatives is crucial. Many employees express a desire to impact their organization's ESG goals but may lack the necessary skills or opportunities. This highlights the importance of investing in learning and development (L&D) programs focused on sustainability training. Empowering employees through education and skill development not only enhances their capabilities but also accelerates progress towards ESG goals.

Furthermore, setting inclusive goals with input from the entire workforce fosters a sense of ownership and joint responsibility for sustainability outcomes. This collaborative approach not only leverages the knowledge and passion of employees but also ensures clear communication and accountability for achieving sustainability targets. Establishing short-term milestones or incremental goals aligned with overarching sustainability objectives can also motivate employees by providing visible progress and recognition for their efforts.

In short, integrating ESG values into the workplace and empowering employees to contribute to sustainability goals are crucial strategies for attracting, engaging, and retaining talent. Organizations that prioritize ESG initiatives and involve employees in goal-setting and implementation will not only see faster progress on sustainability objectives but also experience higher employee satisfaction, engagement, and long-term retention.

Key findings derived from MSCI's ESG data emphasize that ESG performance is not just a checkbox for corporate responsibility but a strategic imperative that directly impacts workforce sentiment, talent attraction, and retention. [xlii]

The study reveals that companies rated as top employers, characterized by high employee satisfaction and attractiveness to talent, consistently exhibit superior ESG scores compared to their industry peers. While their robust environmental performance contributes significantly to this trend, the correlation extends across social and governance aspects as well. This correlation implies that strong ESG performance not only enhances employee satisfaction but also serves as a magnet for prospective talent.

This insight is particularly impactful given previous research indicating that contented employees demonstrate increased productivity, longer tenures, and a greater commitment to achieving organizational goals. Additionally, attracting enthusiastic prospective employees bolsters a company's talent pipeline and secures essential human capital for sustained success.

Looking ahead, the study underscores the escalating importance of ESG performance in talent acquisition and retention strategies, especially with Millennials and Gen Z constituting an increasing proportion of the global workforce. By 2029, these generations are projected to comprise 72% of the workforce, up from 52% in 2019. Importantly, Millennials and Gen Z prioritize environmental and social considerations more than previous cohorts, signaling their heightened expectations from employers regarding ESG commitments.

Consequently, organizations must align their ESG initiatives with these evolving workforce values to remain competitive in attracting and retaining top talent. Failure to do so risks alienating a significant portion of the workforce and losing out on the advantages that robust ESG performance can bring in terms of employee satisfaction, productivity, and overall organizational success.

- **Internal Advocates:**

Employees are not just the workforce; they are the lifeblood of an organization. Engaging them as internal advocates for sustainability initiatives is paramount. Their commitment can turn routine tasks into meaningful environmental and social responsibility contributions.

Establishing Employee Resource Groups (ERGs) focused on sustainability fosters a sense of community and shared purpose, allowing employees to contribute to the organization's sustainability journey actively.

- **Feedback Mechanisms:**

Regular and structured feedback mechanisms, such as surveys, focus groups, and suggestion boxes, create a channel for employees to express their perspectives on sustainability practices. This open communication ensures that their insights influence decision-making processes.

2. Suppliers:

- **Supply Chain Transparency:**

Suppliers play a pivotal role in the sustainability chain. Identifying key suppliers and fostering transparency in the supply chain are essential steps. This involves clearly communicating sustainability expectations, ensuring responsible sourcing, and collaborating to reduce environmental impact.

Creating supplier forums or councils facilitates ongoing dialogue, allowing for the exchange of best practices, addressing challenges, and collectively driving sustainability improvements.

Also, emphasizing sustainability in procurement policies and frameworks is a helpful tool.

- **Capacity Building:**

Recognizing suppliers as strategic partners involves more than contractual agreements. It includes providing resources, training, and support to build their capacity for sustainable practices. This collaborative approach strengthens the overall resilience of the supply chain.

3. Customers:

Consumer preferences are evolving, and there's a clear trend towards prioritizing environmentally and socially responsible products in the consumer packaged goods (CPG) sector. This shift is substantiated by a joint study from McKinsey and NielsenIQ, which delves into the correlation between ESG-related claims on products and actual consumer spending behaviors.

Key Insights from studies related to consumers[xliii]

Consumers exhibit a strong inclination towards products with ESG-related claims, as indicated by their spending patterns over the past five years. Products making such claims saw an average cumulative growth of 28%, outperforming those without claims, which had a growth rate of 20%. This growth is significant, given the competitive nature of the CPG industry.

The study reflects a tangible link between consumers' stated preferences for ESG and their actual purchasing decisions. Over the analyzed period, products with ESG-related claims accounted for 56% of all growth in the categories studied, highlighting a substantial shift in consumer spending habits towards sustainability.

Both large and small brands experienced growth in products featuring ESG-related claims. Smaller brands, in particular, achieved disproportionate growth in 59% of the categories studied, indicating that sustainability can be a key differentiator for emerging players.

The performance of ESG-related claims varied across categories. While certain categories, like sports drinks and hair care, favored smaller brands, larger brands performed better in others, such as fruit juice and sweet snacks, showcasing the nuanced dynamics within each product segment.

Products displaying multiple ESG-related claims demonstrated accelerated growth, suggesting that combining various sustainability aspects can resonate more strongly with consumers. In nearly 80% of categories, products with multiple claims outpaced those with single claims, indicating that consumers perceive authenticity and commitment in brands that address multiple ESG concerns.

The appeal of ESG-related products cuts across demographic boundaries, with various consumer groups showing a preference for such products. While higher-income households and urban residents were more likely to purchase ESG-labelled products, the trend was noticeable across different income levels, life stages, and geographic regions.

In short, the **McKinsey and NielsenIQ** study underscores the growing significance of ESG in consumer purchasing decisions. Companies that proactively integrate sustainability into their product offerings and effectively communicate ESG-related benefits are well-positioned to capitalize on shifting consumer preferences, drive growth, and foster long-term brand loyalty in the evolving CPG landscape.

A recent study conducted by **Bain & Company** sheds light on the evolving landscape of consumer preferences in sustainable consumption.[xliv] The study, which surveyed over 23,000 consumers globally, reveals a growing concern among consumers regarding environmental sustainability and climate change.

The study highlights that nearly two-thirds of respondents globally express "very or extremely" deep concern about environmental sustainability, with climate change concerns intensifying over the past two years.

Fast-growing markets, such as India, Brazil, and China, exhibit higher levels of environmental concern compared to developed countries like the U.S., Germany, and the UK.

Consumers globally indicate a willingness to pay a 12% premium on average for products with minimal environmental impact, with varying degrees of acceptance across different markets.

- **Understanding Preferences:**

Identifying key customer segments and understanding their sustainability preferences are pivotal for customer-centric sustainability strategies. Conducting market research surveys and leveraging data analytics help uncover valuable insights into customer expectations.

Creating customer advisory boards or focus groups provides a direct platform for customers to express their sustainability expectations and preferences.

- **Communication Platforms:**

Engaging customers through various communication channels, including social media, newsletters, and dedicated sustainability reports, keeps them informed. These platforms also offer opportunities for dialogue, enabling organizations to address customer inquiries and share progress transparently.

4. Investors:

There are multifaceted benefits of ESG investing for investors, including financial outperformance, market growth potential, risk management, and alignment with sustainable market trends. As ESG continues to gain prominence in the investment landscape, investors are increasingly recognizing the value and opportunities offered by integrating ESG principles into their portfolios. **Avendus Capital's** analysis underscores the growing significance of ESG factors in investment decisions, highlighting several key reasons why ESG is crucial for investors.[xlv]

Their findings reveal that ESG-focused companies consistently outperform their non-ESG counterparts. This outperformance is a compelling indicator for investors, showcasing the potential for superior financial returns associated with ESG-aligned investments.

The exponential growth of ESG-focused equity funds in India, from $330 million in 2019 to $1.3 billion in June 2023, reflects increasing investor interest and confidence in ESG strategies. Such growth trends indicate a shifting investment landscape towards sustainable and responsible practices.

Avendus Capital predicts a remarkable growth trajectory for ESG funds, with assets under management (AUM) expected to grow by approximately 30% annually over the next 5–10 years. This projection suggests substantial opportunities for investors to capitalize on the evolving ESG market dynamics.

The performance of ESG indices, such as the Nifty 100 ESG index, consistently outperforming traditional benchmarks like the Nifty 100, demonstrates the financial viability and attractiveness of ESG-focused investments. Investors seeking competitive returns are increasingly drawn towards ESG-aligned opportunities.

India's issuance of $3 billion in green bonds in FY24 signifies a growing trend towards sustainable finance, which is expected to foster the development of more green funds in the country. Additionally, the transition to clean energy presents significant economic prospects, leading to the reclassification of Indian companies as ESG assets and creating substantial investment opportunities.

ESG considerations are integral to assessing long-term sustainability and mitigating risks associated with environmental, social, and governance challenges. Investors recognize the importance of incorporating ESG factors into their investment strategies to enhance resilience and ensure sustainable growth.

- **ESG Reporting:**

Investors increasingly consider ESG factors when making investment decisions. Identifying key investors and ensuring transparent ESG reporting is crucial. This involves providing comprehensive information on sustainability initiatives, performance metrics, and future plans.

Regularly updating investors through stakeholder calls, webinars, and annual sustainability reports maintains transparency and aligns the organization's financial objectives with its commitment to responsible practices. ESG Reporting is further detailed in Chapter 7.

5. Local Communities:

- **Understanding Local Needs:**

Local communities are stakeholders with unique needs and concerns. Understanding their priorities requires community engagement initiatives like town hall meetings, community forums, and ethnographic research.

Collaborating with local leaders, community organizations, and NGOs helps identify key issues and ensures that sustainability initiatives resonate with the specific context of each community.

- **Partnerships & Collaboration:**

Forming partnerships with local organizations and community leaders establishes a collaborative approach to sustainability. Joint initiatives address localized challenges and contribute to the well-being of communities surrounding the organization's operations.

6. Regulators & Authorities:

Regulators and authorities emerge as crucial partners, influencing the tempo and trajectory of this transformative journey. Their role is multifaceted, setting the tone, pace, and expectations for organizations engaging in sustainable practices.

In recent years, the landscape of sustainability-related regulations has experienced a significant transformation, with a surge in new directives aimed at enhancing corporate transparency and accountability in ESG matters. The Corporate Sustainability Reporting Directive (CSRD), a notable piece of legislation from the European Union (EU), exemplifies this trend and heralds a new era of stringent reporting standards for large companies operating within the EU.[xlvi]

The CSRD, adopted in November 2022 following substantial support in the European Parliament and Council, marks a substantial shift towards comprehensive sustainability reporting. This directive mandates that large companies in the EU disclose detailed data regarding the impact of their activities on various aspects, including environmental sustainability, social affairs, and governance practices. The ultimate goal is to make transparency on these matters the norm for large firms, reflecting a broader global trend towards sustainable and responsible business practices.

The introduction of the CSRD signifies not just a singular regulatory change but part of a broader movement towards more stringent ESG reporting frameworks. It addresses shortcomings observed in previous directives, such as the Non-Financial Reporting Directive (NFRD), by introducing more detailed reporting requirements aligned with the EU's climate goals and the European Green Deal. This evolution is further fueled by the increasing integration of ESG considerations into financial metrics and performance evaluations.

Moreover, the CSRD's adoption of the European Sustainability Reporting Standards (ESRS) represents a standardized approach to sustainability reporting, aiming to reduce reporting costs and streamline reporting practices across companies. These standards emphasize a double materiality component, requiring companies to report not only on their sustainable activities but also on sustainability activities affecting them.

Looking ahead, the regulatory landscape for sustainability reporting is expected to witness continued evolution and expansion. The CSRD's implementation will impact a significantly larger number of companies, with a nearly fivefold increase in the number of affected entities. This growth underscores the growing emphasis on ESG considerations and the need for companies to align with stringent reporting standards to provide reliable and comparable data to investors and stakeholders.

Regulators as Architects of Direction:

Regulators function as architects, designing the blueprint for sustainability within industries and sectors. The regulations they formulate set the direction, guiding organizations on how fast or slowly they must navigate towards sustainable practices. These regulations serve as the compass, indicating the path that aligns with broader ESG objectives.

- **Defining Standards:** Regulatory frameworks establish standards and benchmarks, providing a common language for sustainable practices. Whether it's emission limits, waste disposal guidelines, or social responsibility mandates, these standards offer a clear trajectory for organizations to follow.
- **Influencing Corporate Behavior:** The regulatory environment shapes corporate behavior by delineating acceptable boundaries and encouraging proactive engagement in sustainable practices. Organizations, cognizant of regulatory expectations, often tailor their strategies to not only comply but excel in sustainability.

Impact on Investments in Sustainability:

Regulatory frameworks wield considerable influence over the financial landscape, impacting investments in sustainability. The nature and stringency of regulations can significantly affect how organizations allocate resources and prioritize sustainability initiatives.

- **Creating Financial Incentives:** Governments and regulatory bodies often introduce financial incentives to encourage sustainability. Tax credits, grants, or subsidies for eco-friendly practices can drive organizations to invest more heavily in sustainable technologies and processes.

- **Risk Mitigation & Investor Confidence:** Stringent regulations can mitigate the risks associated with non-compliance, bolstering investor confidence. Investors, increasingly attuned to ESG considerations, view adherence to regulations as a sign of robust governance and risk management, making sustainability-focused investments more attractive.
- **Shift In Capital Allocation:** Regulatory pressures may lead to a shift in capital allocation. As regulations evolve, organizations may divest from non-sustainable ventures and redirect investments towards initiatives aligned with regulatory expectations.

The Dynamic Regulatory Landscape:

The regulatory landscape is dynamic, evolving in response to global challenges, public opinion, and emerging sustainability trends. Keeping pace with this dynamism requires organizations to not only comply with existing regulations but also anticipate and adapt to future changes.

- **Adaptability as a Competitive Advantage:** Organizations that demonstrate adaptability to evolving regulatory landscapes gain a competitive edge. Proactively aligning strategies with anticipated regulatory shifts positions them as leaders in sustainability, well-prepared for the challenges and opportunities that lie ahead.
- **Stakeholder Expectations & Regulatory Alignment:** In an era of heightened stakeholder expectations, regulatory alignment becomes a strategic imperative. Organizations responsive to the broader societal push towards sustainability can navigate regulatory changes more smoothly, minimizing disruptions to their operations.

In the context of sustainability, regulators and authorities wield considerable influence in shaping the direction and scope of sustainable practices. Through their regulatory frameworks, these entities establish the standards and expectations that guide the pace and rhythm of sustainability initiatives. For organizations, a thorough understanding and effective navigation of this regulatory landscape is not a mere compliance exercise but rather a strategic imperative that determines the trajectory of their future operations. As regulatory frameworks continue to evolve and expand, organizations that are able to harmonize with these standards and expectations will not only achieve compliance but also find themselves in concert with the evolving demands of a world that is increasingly focused on sustainability.

ESG Rating:

In recent years, ESG rating agencies have emerged as ESG considerations have gained significant importance in the corporate world as stakeholders increasingly demand transparency and accountability from companies. ESG rating agencies play a crucial role in assessing and benchmarking companies' performance in these areas, providing valuable insights for investors, consumers, and other stakeholders.

ESG rating agencies are organizations that evaluate and rate companies based on their performance in environmental, social, and governance practices. These agencies collect data from various sources, including company disclosures, public records, and third-party providers, to assess how well companies are managing ESG risks and opportunities. They then assign a rating or score to each company, indicating its ESG performance relative to peers or industry benchmarks. The agencies employ a variety of methodologies to evaluate companies' ESG performance. They typically look at factors such as carbon emissions, diversity and inclusion practices, executive compensation, board diversity, human rights policies, supply chain management, and more. Some agencies use quantitative metrics, while others rely on qualitative analysis and expert judgment to assess ESG performance. Companies are often required to provide detailed information about their ESG practices and performance to the rating agencies, which may include submitting reports, surveys, and other documentation. Some agencies also engage with companies directly through interviews or site visits to better understand their ESG initiatives.

ESG rating agencies play a crucial role in promoting transparency, accountability, and sustainability in the corporate world. Investors use ESG ratings to assess companies' long-term performance and risk exposure, integrating ESG factors into their investment decisions. Consumers and other stakeholders also rely on ESG ratings to make informed choices and hold companies accountable for their social and environmental impacts. By providing standardized ESG ratings and benchmarks, these agencies help companies identify areas for improvement, set targets, and track progress over time. ESG ratings can also enhance companies' reputations, attract investors, and improve access to capital by demonstrating a commitment to sustainability and responsible business practices.

Companies should internally asses if they want to seek an ESG Rating. Some agencies will rate you without your consent based on public information. If a company decide to seek an ESG rating, it should consider and expand several key factors to ensure a meaningful assessment of its performance:

1. Transparency and Disclosure: Companies should provide accurate, timely, and comprehensive information about their ESG practices and performance to the rating agencies. Transparent disclosure helps build trust with stakeholders and enhances the credibility of the ESG rating. More details are in chapter 9.

2. Materiality: Companies should focus on ESG issues that are most relevant to their business and stakeholders. Identifying material ESG risks and opportunities ensures that companies prioritize actions that create value and mitigate risks effectively. More details are in Chapter 2.

3. Integration into Business Strategy: ESG considerations should be integrated into the company's overall business strategy and decision-making processes. Companies should demonstrate a commitment to sustainability and responsible practices throughout their operations. More details in chapter 6.

4. Continuous Improvement: ESG ratings are not static and can change over time based on companies' performance and external factors. Companies should set measurable ESG targets, implement initiatives to improve performance, and regularly monitor and report progress. More details in chapter 8.

5. Engagement with Stakeholders: Companies should engage with investors, customers, employees, and other stakeholders to understand their expectations and concerns regarding ESG issues. Meaningful stakeholder engagement can help companies address key ESG challenges and build stronger relationships with their stakeholders. More details are in chapter 4.

Overall, ESG rating agencies play a critical role in assessing and benchmarking companies' performance on ESG factors and make their performance comparable for all relevant stakeholders.

Communication is vital for managing stakeholders

Open and transparent communication is the linchpin that not only keeps stakeholders informed about the strategy's evolution but also cultivates trust and understanding.

Actively engaging and managing stakeholders is a critical aspect of successful sustainability initiatives. To foster meaningful connections, organizations must adopt a proactive approach that transcends mere communication.

This involves identifying and understanding the diverse needs, expectations, and concerns of stakeholders. Open and transparent communication channels should be established, providing stakeholders with a platform to express their perspectives. Engaging in genuine dialogue, listening attentively, and incorporating stakeholder feedback into decision-making processes are fundamental practices.

Additionally, organizations should leverage technology and various communication channels to disseminate information effectively. Moreover, establishing formal mechanisms for stakeholder involvement, such as advisory groups or forums, enhances the inclusivity of the engagement process.

The key lies not only in disseminating information but also in creating a collaborative environment where stakeholders feel valued, heard, and integral to the sustainability journey. Regularly assessing the impact of engagement efforts and adapting strategies based on feedback further solidifies the organization's commitment to building lasting and mutually beneficial relationships with its stakeholders.

Open & Transparent Communication	Actively Listening, Collaborating, & Adapting
•Open and Transparent Communication	•Actively Listening To Stakeholders' Concerns:
•Active Stakeholder Involvement In Decision-Making	•Offering Opportunities For Stakeholder Participation
•Building Strong Relationships	•Acknowledging & Appreciating Stakeholder Contributions
•Understanding Motivations, Values, & Concerns	•Continuous Evaluation & Adaptation

Open & Transparent Communication:

- **Regular Updates & Reports**

Providing regular updates and reports on the development and progress of the sustainability strategy is foundational. This includes detailed documentation, such as sustainability reports, progress dashboards, and newsletters, that stakeholders can access easily.

Clear and concise language in communication materials ensures that technical details are accessible to a broad audience, fostering inclusivity in understanding the strategy's objectives and milestones.

- **Interactive Platforms**

Leveraging interactive platforms, such as webinars, town hall meetings, and Q&A sessions, facilitates direct communication. These platforms allow stakeholders to pose questions, seek clarifications, and actively engage with the organization's leadership and sustainability teams.

Utilizing digital tools, including social media, online forums, and dedicated communication channels, creates an environment where stakeholders can share feedback and discuss strategy-related matters in real time.

Active Stakeholder Involvement in Decision-Making:

- **Stakeholder Advisory Boards**

Establishing stakeholder advisory boards or committees provides a structured mechanism for involving stakeholders in decision-making. These boards can represent diverse stakeholder groups and contribute insights on strategy development, implementation, and impact assessment.

Actively seeking input on specific decisions, such as setting sustainability goals or choosing impactful initiatives, ensures that the perspectives of various stakeholders are considered, fostering a sense of shared ownership.

- **Collaborative Workshops & Consultations**

Organizing collaborative workshops and consultations invites stakeholders into the decision-making process. These sessions can focus on key strategy elements, inviting stakeholders to share their expertise, expectations, and concerns.

Platforms for virtual or in-person engagement, where stakeholders can provide real-time input on critical decisions, not only enhance the quality of decision-making but also demonstrate a commitment to inclusivity.

Building Strong Relationships:

- **Stakeholder Mapping & Analysis**

Conducting thorough stakeholder mapping and analysis helps identify key individuals or groups. Understanding the dynamics of each stakeholder's interests, motivations, and concerns lays the groundwork for tailored communication and relationship-building.

Utilizing stakeholder personas based on demographic, geographic, or industry-specific factors allows for a nuanced approach to understanding and addressing the unique needs of different stakeholder segments.

- **Individual & Group Meetings**

Investing time in individual and group meetings with key stakeholders builds personal connections. These interactions provide opportunities to delve deeper into stakeholder motivations, values, and concerns, fostering a deeper understanding of their perspectives.

Engaging in one-on-one conversations with influential stakeholders, such as community leaders or prominent customers, not only strengthens relationships but also allows for more targeted and impactful communication.

Understanding Motivations, Values, & Concerns:

- **Stakeholder Surveys & Feedback Sessions**

Conducting stakeholder surveys and feedback sessions helps organizations glean insights into stakeholder motivations, values, and concerns. Structured surveys can include questions about stakeholder expectations, the perceived impact of the sustainability strategy, and areas for improvement.

Actively seeking feedback on specific aspects of the strategy, such as the organization's approach to diversity and inclusion or environmental stewardship, demonstrates a commitment to continuous improvement based on stakeholder input.

- **Collaborative Goal-Setting**

Collaboratively setting sustainability goals with stakeholders ensures alignment with their values and expectations. This involves soliciting input on the prioritization of goals, their specific metrics, and the desired outcomes.

Integrating stakeholder-driven goals into the overall sustainability strategy not only reflects a responsive approach but also increases the likelihood of successful goal achievement through shared commitment.

Open and transparent communication, active stakeholder involvement in decision-making, and investing in strong relationships constitute the pillars of successful stakeholder engagement. By embracing these practices, organizations can build a foundation of trust, foster a sense of shared responsibility, and ensure that the sustainability strategy is not merely a document but a living, evolving commitment.

For example,

Unilever, consumer goods giant, is known for its commitment to sustainability and open communication. Through its Sustainable Living Plan, Unilever sets clear goals for reducing environmental impact and improving social conditions. The company regularly communicates progress through detailed sustainability reports and actively engages with stakeholders through initiatives like the Sustainable Brands platform. Unilever's transparency has not only built trust but has also encouraged other companies to embrace sustainable practices.[xlvii]

Patagonia, an outdoor clothing company, is a trailblazer in active stakeholder involvement. The company actively engages its customers in decision-making processes related to environmental causes. For instance, the "Worn Wear" initiative encourages customers to buy used Patagonia gear or trade in their old items. This not only involves customers in the circular economy but also aligns with their values of environmental responsibility. Patagonia's engagement strategies have turned customers into advocates for sustainability.[xlviii]

Microsoft has demonstrated a commitment to building strong relationships with various stakeholders. The company engages in inclusive stakeholder forums, seeking input from customers, employees, and industry experts. Microsoft's consistent efforts to understand the values and concerns of stakeholders are reflected in initiatives such as the AI for Accessibility program, which actively involves people with disabilities in the development process. This approach not only builds relationships but also ensures that products are inclusive and address diverse needs.[xlix]

Danone, a multinational food products corporation, actively collaborates with stakeholders to understand their motivations and values. Through initiatives like the Danone Ecosystem Fund, the company engages with local communities and entrepreneurs to address social and environmental challenges. This collaborative approach ensures that Danone's sustainability strategy aligns with each stakeholder group's unique motivations and values. By understanding local needs, Danone has built meaningful relationships that go beyond business transactions.[l]

IKEA, a global furniture retailer, emphasizes understanding stakeholder motivations through surveys and feedback sessions. The company regularly conducts surveys to gather insights from customers, employees, and suppliers. Additionally, IKEA engages in co-creation projects, involving customers in the design and development of sustainable products. By actively seeking input and involving stakeholders in decision-making, IKEA ensures that its sustainability initiatives resonate with the values and expectations of its diverse stakeholder groups.[li]

These examples showcase how organizations across different industries have leveraged open communication, active stakeholder involvement, and relationship-building to drive transformative sustainability initiatives. Each of these companies has demonstrated that engaging stakeholders goes beyond meeting expectations—it is about co-creating a sustainable future together.

Actively Listening, Collaborating, & Adapting

Actively listening to stakeholders, fostering collaboration, and adapting strategies based on feedback are essential practices. This section explores how organizations can navigate concerns, involve stakeholders in implementation, express appreciation, and continually refine their stakeholder engagement efforts.

Actively Listening to Stakeholders' Concerns:

- **Open Dialogue Platforms**

Establishing open dialogue platforms, such as regular feedback sessions, town hall meetings, or dedicated helplines, allows stakeholders to voice concerns. Actively listening to these concerns demonstrates a commitment to understanding and addressing issues promptly.

Utilizing digital platforms, surveys, and social media channels for real-time interaction ensures that concerns are captured in a timely manner and responses are agile.

- **Conflict Resolution Strategies**

Developing conflict resolution strategies is crucial for addressing disagreements constructively. Creating a designated conflict resolution team or mediator role can help in facilitating discussions, finding common ground, and reaching mutually beneficial solutions.

Emphasizing transparency in the resolution process builds trust, and organizations can use lessons learned from conflicts to strengthen future stakeholder relationships.

Involving industry federations and relevant establishments further enriches the resolution process, contributing to the depth of transparency and trust-building. Collaborating with industry federations provides a broader perspective, aligning conflict resolution with industry standards and best practices. These partnerships bring collective wisdom to the table, fostering a more comprehensive understanding of stakeholders' concerns.

Furthermore, engaging with industry establishments allows organizations to tap into shared experiences and knowledge, leveraging lessons learned from similar conflicts across the sector.

This collaborative approach not only enhances the resolution process but also fortifies future stakeholder relationships by instilling a sense of industry-wide commitment to transparency and continuous improvement. As organizations learn from conflicts, the involvement of industry federations becomes a strategic asset, reinforcing a culture of openness and cooperation within the broader professional ecosystem.

Offering Opportunities for Stakeholder Participation:

- **Advisory Roles & Focus Groups**

Offering stakeholders advisory roles, such as participation in sustainability committees or focus groups, provides a platform for deeper involvement. This collaborative approach ensures that stakeholders contribute to decision-making processes.

Conducting regular focus group sessions for specific initiatives, where stakeholders can share insights and co-create solutions, fosters a sense of ownership and strengthens the quality of strategic decisions.

- **Collaborative Project Initiatives**

Involving stakeholders in collaborative project initiatives related to the strategy's implementation creates a sense of shared responsibility. This could include joint sustainability projects, community engagement programs, or the co-development of sustainable products.

Establishing cross-functional teams that include both internal and external stakeholders ensures diverse perspectives and expertise are brought to the table.

Acknowledging & Appreciating Stakeholder Contributions:

- **Recognition Programs**

Implementing recognition programs acknowledges and appreciates the contributions of stakeholders. This could involve highlighting successful sustainability initiatives led by stakeholders, showcasing their impact, and publicly expressing gratitude.

Recognizing stakeholders in sustainability reports, social media platforms, or at public events not only strengthens relationships but also inspires others to engage in the organization's sustainability journey actively.

- **Customized Appreciation**

Tailoring appreciation efforts based on stakeholder preferences adds a personal touch. For example, sending personalized thank-you notes, organizing appreciation events, or offering exclusive benefits as a token of gratitude.

Understanding the cultural and professional norms of different stakeholder groups ensures that appreciation efforts are received positively and resonate with the values of each group.

Continuous Evaluation & Adaptation:

- **Feedback Mechanisms**

Establishing systematic feedback mechanisms is fundamental to continuous evaluation. Regular surveys, post-engagement evaluations, and feedback forms provide stakeholders with a structured way to share their perspectives on the effectiveness of engagement efforts.

Encouraging stakeholders to provide constructive feedback on the engagement process itself ensures that the organization remains responsive to evolving expectations.

- **Adaptive Strategies**

Being willing to adapt engagement strategies based on feedback and changing circumstances is a hallmark of successful stakeholder engagement. This adaptability may involve refining communication channels, adjusting the frequency of engagement activities, or introducing new initiatives.

Conducting periodic reviews of stakeholder engagement strategies and incorporating lessons learned into future plans ensures that the organization remains agile and responsive.

Actively listening, collaborating, and adapting represent the heartbeat of sustainable stakeholder engagement. Organizations can cultivate enduring relationships that transcend transactional interactions by fostering an environment where concerns are heard, collaboration is encouraged, and recognition is genuine.

Conclusion

Effective stakeholder engagement is the cornerstone of a sustainable and responsible business strategy. By recognizing and involving key internal and external stakeholders, organizations gain valuable insights, build trust, and ensure that their sustainability initiatives align with diverse expectations. From engaging employees as internal advocates to fostering transparent relationships with suppliers, customers, investors, and local communities, the collaborative efforts create a robust foundation for meaningful sustainability progress. Furthermore, the pivotal role of regulators and authorities in shaping the trajectory of sustainable practices underscores the broader societal impact.

Organizations must not only comply with existing regulations but also proactively adapt to evolving standards, positioning themselves as leaders in sustainability. Open and transparent communication, active stakeholder involvement, and continuous adaptation emerge as fundamental practices, fostering a shared commitment to building a sustainable future together.

The success stories of companies like Unilever, Patagonia, Microsoft, Danone, and IKEA serve as inspirations, showcasing the transformative power of stakeholder engagement in driving sustainable business practices and positively impacting the world.

Chapter 5
Develop an ESG Strategy

"Developing an ESG strategy isn't just a checkbox—it's a pathway to long-term resilience, positive impact, and ethical leadership."

A comprehensive strategy is crucial for an organization to adopt sustainable and responsible practices. This strategy should be developed after thoroughly assessing the current practices and defining clear goals. It should address the organization's environmental and social impacts. This chapter explores the key components and considerations involved in crafting such a sustainability strategy.

Assessment & Goal-Setting is the Foundation for Strategy Development

The journey towards sustainability begins with a deep dive into the current state of an organization. Assessing environmental, social, and economic impacts lays the foundation for informed decision-making. This diagnostic phase involves evaluating energy and resource use, waste generation, emissions, supply chain practices, and the overall societal impact of the organization on communities and employees.

Building upon this assessment, the next crucial step is goal-setting. Sustainability goals must be specific, measurable, achievable, relevant, and time-bound (SMART). These goals should resonate with the organization's values and long-term vision and address the most material environmental and social impacts. The goal-setting phase not only charts the course for the organization's sustainability journey but also serves as the guiding star for the subsequent development of a comprehensive strategy.

Components Of a Comprehensive Sustainability Strategy

The journey towards sustainability requires a strategic and holistic approach that encompasses ESG considerations. The foundation of such a comprehensive strategy rests on five key pillars, each addressing specific facets critical to fostering responsible business practices. The five key pillars are environmental initiatives, social initiatives, governance initiatives, innovative technologies, and circular economy principles.

In the world of environmental initiatives, the focus is on optimizing energy usage and minimizing waste through innovative technologies and circular economy principles.

Social initiatives are about organizational culture, promoting diversity, equity, and inclusion while fostering relationships through responsible sourcing practices.

Governance initiatives extend beyond internal operations to community engagement, recognizing the importance of understanding and addressing the needs of local stakeholders.

Innovative technologies highlight the importance of leveraging technological advancements to support environmental goals, such as energy efficiency, waste reduction, and sustainable resource management.

Circular economy principles emphasize the adoption of circular economy principles, including designing products for longevity, promoting reuse and recycling, and minimizing resource depletion throughout the product lifecycle.

Together, these five components form the core elements of a robust sustainability strategy, aligning the organization with the principles of ESG for a positive impact on the environment, society, and corporate governance.

Crafting a Roadmap for Change

With goals in place and a clear understanding of the organization's impact, the development of a sustainability strategy takes shape. This strategy serves as a roadmap, guiding the integration of sustainable practices into core business operations. It becomes a living document that outlines how the organization plans to navigate the complexities of environmental stewardship, social responsibility, and economic viability.

Sustainability is fundamentally about change and transformation. It necessitates a shift in mindset, practices, and systems to align with the principles of responsible business. The importance of establishing a culture of change within the organization cannot be overstated. This cultural shift involves instilling a sense of responsibility, innovation, and adaptability across all levels. It requires employees to embrace new ways of thinking and working, fostering a collective commitment to sustainability.

In the pursuit of sustainability, organizations must always work towards their "north star." This metaphorical guiding principle represents the overarching vision and long-term objectives that shape the sustainability strategy. It serves as a constant reference point, aligning all initiatives and decisions with the ultimate goal of creating positive environmental, social, and economic impact. The North Star provides clarity, purpose, and direction, ensuring that the organization remains on course amidst the dynamic landscape of sustainable practices.

Integration Into Business Operations

A robust sustainability strategy is not a standalone document but an integrated part of the organization's DNA. It permeates through supply chain management, product design, manufacturing processes, and customer engagement. The strategy aligns with business objectives, ensuring that sustainability is not a mere addendum but an inherent driver of operational excellence.

Implementing Environmental (E) & Social (S) Initiatives in Business Operations:

To truly embed sustainability into the organizational fabric, it's essential to implement specific initiatives within day-to-day operations. This involves identifying opportunities to enhance energy efficiency, reduce waste, and foster a culture of environmental and social responsibility across all business functions. Integrating E and S initiatives ensures that every aspect of the organization contributes to its broader sustainability goals.

Remember, implementing environmental (E) and social (S) initiatives in business operations is not just about adhering to regulatory requirements or societal expectations; it's about strategically leveraging these initiatives to create long-term value for the company.

A strong ESG proposition enables companies to tap into new markets and expand their existing ones. For instance, companies with superior ESG execution often gain preferential treatment from governing authorities, leading to easier access, approvals, and licenses for growth opportunities. An example of this can be seen in Long Beach, California, where for-profit companies selected for a massive infrastructure project were screened based on their sustainability performance, showcasing how ESG execution can pay off in terms of market access and growth opportunities.[lii]

Secondly, effective ESG strategies can significantly reduce operational costs. Companies that prioritize sustainability can combat rising expenses, such as raw-material costs and environmental impact costs like water and carbon usage. This not only positively affects operating profits but also enhances resource efficiency and financial performance. For instance, companies like **3M** have saved billions by proactively addressing environmental risks through initiatives like pollution prevention programs, while major water utilities have achieved substantial cost savings through lean initiatives focused on energy consumption and waste reduction.[liii]

Moreover, a strong external value proposition derived from robust ESG practices can lead to greater strategic freedom and reduced regulatory pressures. Companies that excel in ESG are less likely to face adverse government actions and can even garner government support. By aligning with regulatory goals and demonstrating a commitment to sustainability, companies can mitigate the risks associated with regulatory compliance and potential legal interventions.

Furthermore, ESG initiatives contribute to attracting and retaining quality talent, enhancing employee motivation, and ultimately increasing overall productivity. Positive social impact and a sense of purpose within the organization lead to higher job satisfaction and performance. For example, companies listed among Fortune's "100 Best Companies to Work For" have consistently demonstrated higher stock returns, showcasing the link between employee satisfaction, social impact, and shareholder value.[liv]

Lastly, a robust ESG proposition enhances investment returns by directing capital towards sustainable opportunities and avoiding investments with potential long-term environmental risks. Companies that proactively invest in renewable energy, waste reduction, and sustainable practices not only contribute to environmental stewardship but also position themselves for long-term financial gains. For instance, companies like Unilever and Nestle have successfully capitalized on sustainable products and practices, generating substantial profits from renewable and environmentally friendly offerings.

- **Adding Environmental Considerations (E) to Existing Products:**

Existing products can be a significant source of environmental impact. To address this, a sustainable strategy involves analyzing the life cycle of products, identifying areas for improvement, and incorporating environmental considerations. This may include using eco-friendly materials, optimizing packaging for reduced waste, or extending product life through repair and refurbishment programs.

The automotive industry is a prime example of this shift. Automobile manufacturers are increasingly shifting from conventional internal combustion engine vehicles to electric vehicles (EVs). This transition addresses environmental concerns related to air pollution and greenhouse gas emissions. Companies like Tesla, Nissan, and BMW have introduced electric cars that run on battery power, reducing reliance on fossil fuels and lowering carbon emissions over the vehicle's life cycle.[lv]

Many companies are adopting eco-friendly materials in their products to reduce environmental impact. For instance, fashion brands like Patagonia and Stella McCartney use sustainable fabrics such as organic cotton, recycled polyester, and Tencel (made from renewable wood sources) in their clothing lines.[lvi] These materials require fewer resources and have lower environmental footprints compared to traditional textiles.

Consumer goods companies are rethinking their packaging strategies to minimize waste generation. For example, Unilever has committed to making all its plastic packaging recyclable, reusable, or compostable by 2025.[lvii] They are also exploring packaging alternatives like biodegradable materials and lightweight designs to reduce material consumption and waste generation throughout the product life cycle.

Electronics companies are promoting product longevity through repair and refurbishment programs. Apple, for instance, offers its customers the option to repair and upgrade their devices instead of replacing them entirely.[lviii] This not only reduces electronic waste but also extends the useful life of products, contributing to a more sustainable consumption model.

- **Developing New Products With a Focus on Environmental Impact (E):**

Innovation plays a crucial role in sustainability. Developing new products with a primary focus on environmental impact allows organizations to align their offerings with the principles of sustainability. This could involve creating energy-efficient products, made from sustainable materials, or contribute positively to environmental conservation.

Allbirds' M0.0NSHOT shoe is the world's first net-zero carbon shoe, showcasing innovation for sustainability. This groundbreaking product showcases how organizations can develop new offerings with a primary focus on environmental impact.[lix]

The M0.0NSHOT shoe achieves a net-zero carbon footprint, meaning it emits zero carbon dioxide equivalent (CO_2e) during its production and use phases. This is a significant achievement compared to a standard sneaker, which typically has a carbon footprint of about 14 kg CO_2e.

What sets this shoe apart is that it attains this net-zero status without relying on carbon offsets, thanks to SweetFoam® tech, carbon labeling, and partnerships like with Adidas. Using regenerative wool and carbon-negative materials, along with carbon-conscious packaging and transportation, Allbirds sets a new standard for eco-friendly footwear.

- **Embedding Social (S) Responsibility in Processes:**

Sustainability is not solely about environmental considerations; the social dimension is equally vital. Embedding social responsibility in processes involves creating an organizational culture that prioritizes fair labor practices, diversity and inclusion, and community engagement. This could mean implementing ethical sourcing policies, promoting workforce well-being, and actively engaging with local communities.

In conclusion, integrating sustainability into business operations requires a comprehensive approach that addresses both environmental and social aspects. It involves not only implementing specific initiatives but also reimagining existing products and processes to align with sustainability goals. This holistic integration ensures that sustainability becomes an intrinsic part of the organizational DNA, driving operational excellence and responsible business practices.

Continuous Improvement

A commitment to continuous improvement is woven into the fabric of an effective sustainability strategy. Learning from successes and challenges, organizations adapt and refine their approach. Flexibility and adaptability are key as the landscape of sustainability evolves and new opportunities and challenges emerge.

- **Linking Continuous Improvement to Cultural Change & Transformation:**

Sustainability is not just a checklist of initiatives; it's a cultural shift and transformation. Organizations committed to continuous improvement in sustainability understand the need to instill a culture that values innovation, responsibility, and resilience. This cultural change involves fostering a mindset where every member of the organization actively contributes to the sustainability journey. As practices evolve and new insights emerge, the organization adapts, ensuring that sustainability becomes ingrained in its collective identity.

- **Adding Sustainability to the DNA of the Organization:**

Continuous improvement goes beyond periodic adjustments; it involves embedding sustainability into the DNA of the organization. This means integrating sustainable practices into the core values, mission, and day-to-day operations. Sustainability becomes a guiding principle that influences decision-making at every level. From leadership to frontline employees, everyone understands their role in contributing to sustainability goals. This cultural integration ensures that the commitment to continuous improvement becomes an enduring aspect of the organization's identity.

Defining Long, Medium, & Short-Term Ambitions

When it comes to a comprehensive sustainability plan, setting goals for the short and long term is like planning a journey. These goals act as guiding lights, showing the way towards a sustainable and responsible future.

For a deeper understanding of how to establish and align these ambitions effectively, refer to Chapter 3 for more detailed insights and strategies.

- **Why Focus on Short, Medium, & Long-Term Impact:**

    o **Strategic Alignment:** Goals that encompass short, medium, and long-term horizons ensure strategic alignment. Short-term goals may address immediate challenges, while medium- and long-term goals align with the broader vision, fostering consistency and coherence in sustainability efforts.

    o **Adaptability:** The dynamic nature of sustainability requires organizations to be adaptable. Short-term goals allow for quick wins and adjustments, while long-term goals provide a visionary framework. The medium-term serves as a bridge, allowing organizations to adapt strategies based on evolving circumstances and emerging trends.

    o **Measurable Progress:** The tiered structure of short-, medium-, and long-term goals facilitates measurable progress. Short-term goals offer tangible milestones; medium-term goals assess ongoing initiatives; and long-term goals provide a holistic measure of the organization's transformative journey.

    o **Comprehensive Impact:** Short-term goals address immediate operational improvements; medium-term goals focus on embedding sustainability into core processes; and long-term goals drive systemic change. This comprehensive approach ensures that sustainability permeates every facet of the organization's operations.

Prioritization of Initiatives Under The E, S & G

In the complex field of sustainability strategy, prioritizing initiatives under the pillars of Environmental, Social, and Governance (ESG) is a strategic endeavor that must balance impact, feasibility, and alignment with organizational values.

Link of ESG Strategy to Sustainable Development Goals

In the wake of global challenges such as climate change, social inequality, and environmental degradation, the **United Nations** introduced the Sustainable Development Goals (SDGs) in 2015.

Goal 1: No Poverty

Goal 2: Zero hunger

Goal 3: Good health and well-being

Goal 4: Quality education

Goal 5: Gender equality

Goal 6: Clean water and sanitation

Goal 7: Affordable and clean energy

Goal 8: Decent work and economic growth

Goal 9: Industry, Innovation, and Infrastructure

Goal 10: Reduced inequality

Goal 11: Sustainable cities and communities

Goal 12: Responsible consumption and production

Goal 13: Climate action

Goal 14: Life below water

Goal 15: Life on land

Goal 16: Peace, justice, and strong institutions

Goal 17: Strengthen the means of implementation and revitalize the global partnership for sustainable development goals.

These 17 interconnected goals serve as a blueprint for a better and more sustainable future for all, aiming to address pressing issues facing our world by 2030[lx]. While the SDGs were initially created as a guide for governments, their relevance extends far beyond the public sector. Businesses, both large corporations and small enterprises alike, play a crucial role in achieving the SDGs and promoting sustainable development.

Businesses are increasingly recognizing the importance of aligning their strategies with the SDGs.

There are several reasons why understanding and integrating the SDGs into corporate strategies are essential for companies in today's world:

- **Risk Mitigation and Resilience:**

Embracing the SDGs helps companies identify and mitigate risks associated with environmental, social, and governance issues. By aligning with the goals, businesses can build resilience to future challenges, such as regulatory changes, resource scarcity, and shifting consumer preferences.

- **Enhanced Reputation and Stakeholder Engagement:**

Companies that actively support the SDGs enhance their reputation as responsible corporate citizens. Engaging with the goals can strengthen relationships with stakeholders, including customers, investors, employees, and communities, leading to increased trust and loyalty.

- **Market Opportunities and Competitive Advantage:**

The SDGs present numerous market opportunities for innovative products, services, and business models. By aligning with the goals, companies can gain a competitive edge, attract new customers, and access new markets while driving sustainable growth and profitability.

- **Cost Savings and Efficiency:**

Adopting sustainable practices in line with the SDGs can lead to cost savings through improved resource efficiency, waste reduction, and energy conservation. Companies that prioritize sustainability often see long-term financial benefits and operational efficiencies.

- **Attracting and Retaining Talent:**

In today's competitive job market, employees increasingly seek purpose-driven organizations that are committed to making a positive impact. Aligning with the SDGs can help companies attract top talent, foster employee engagement, and improve retention rates.

- **Regulatory Compliance and License to Operate:**

Governments around the world are increasingly incorporating the principles of the SDGs into policies and regulations. By aligning with the goals, companies can ensure compliance with evolving regulatory frameworks and maintain their social license to operate.

- **Long-Term Value Creation:**

Embedding the SDGs into corporate strategies fosters long-term value creation by considering environmental, social, and governance factors alongside financial performance. Sustainable businesses are better positioned to adapt to changing market dynamics and create enduring value for all stakeholders. In conclusion, the Sustainable Development Goals offer a compelling framework for businesses to contribute to a more sustainable and inclusive future while driving innovation, growth, and resilience.

Companies that understand the relevance of the SDGs and align their strategies accordingly are not only better positioned to thrive in a rapidly changing world but also play a crucial role in advancing sustainable development on a global scale. Embracing the SDGs is not just a moral imperative but a strategic imperative for businesses seeking to create lasting positive impact and secure their place in a more sustainable future.

Link Of ESG Strategy to the Corporate Strategy

By ensuring a direct link between the ESG strategy and the corporate strategy, organizations create a synergistic approach that aligns progress with the overarching purpose of the business.

To maximize impact, the ESG strategy should seamlessly align with the broader corporate strategy. This alignment ensures that sustainability goals are embedded in the core business objectives, creating a unified approach to organizational success.

If the corporate strategy emphasizes innovation, the ESG strategy might align by prioritizing initiatives that promote sustainable product design, eco-friendly packaging, and circular economy practices.

To fully integrate sustainability, organizations should incorporate ESG considerations into decision-making at all levels. This ensures that sustainability is not a standalone initiative but an integral factor in shaping business strategies.

During product development, incorporating ESG considerations might involve assessing the environmental impact of materials, evaluating social implications in the supply chain, and ensuring ethical governance in sourcing.

Transparent communication about how the ESG strategy contributes to the broader corporate strategy fosters stakeholder understanding and support. It emphasizes that sustainability is not just a responsibility but a strategic driver of long-term success.

Through annual reports, corporate communications, and stakeholder engagements, organizations can articulate how achieving sustainability targets directly contributes to the overall resilience, reputation, and financial success of the business.

Continuous Improvement Is a Dynamic Approach to ESG

Establishing feedback loops ensures that insights gained from the monitoring of yearly targets and KPIs are used for continuous learning. This iterative process enables organizations to refine strategies and enhance the effectiveness of sustainability initiatives.

Conducting regular stakeholder surveys and employee feedback sessions provides valuable insights into the perceived impact of sustainability efforts and potential areas for improvement.

The ESG landscape is dynamic, with new challenges and opportunities continually emerging. An adaptive approach ensures that the ESG strategy remains responsive to evolving trends, regulations, and stakeholder expectations.

If new regulations that impact the organization's environmental practices are introduced, the ESG strategy should be flexible enough to incorporate these changes while maintaining alignment with the corporate strategy.

What Resources/Investments Are Required To Achieve Your Strategy?

To achieve our sustainability goals, we must make strategic investments and allocate resources to turn our aspirations into reality. Investments in financial, human, technological, and collaborative resources are essential for cultivating a sustainable future. By strategically allocating resources, organizations not only meet their ESG goals but also create a foundation for long-term prosperity that balances environmental responsibility, social impact, and ethical governance.

1. Financial Investments- Sowing Seeds for Sustainability	2. Human Capital- Nurturing the Seeds of Change	3. Technological Infrastructure- Building the Foundation for Change	4. Stakeholder Engagement- Cultivating Partnerships for Impact
• Renewable Energy Infrastructure	• Dedicated Sustainability Teams:	• Environmental Monitoring Technologies:	• Community Partnership Programs:
• Research and Development (R&D):	• Skilled Talent Acquisition:	• Data Analytics and Reporting Tools:	The organization's social license to operate. • Collaboration with NGOs and Industry Groups:
• Training and Education Programs:	• External Consultants and Advisors:	• Supply Chain Management Systems:	• Regular Stakeholder Communication Platforms:

1. Financial Investments- Sowing Seeds for Sustainability

Funding for the installation of renewable energy infrastructure, such as solar panels or wind turbines, to transition towards cleaner and more sustainable energy sources.

Reducing reliance on fossil fuels and embracing renewable energy aligns with environmental goals and contributes to long-term cost savings.

Allocation of funds for R&D initiatives focused on sustainable product development, eco-friendly packaging, and innovative technologies that reduce environmental impact.

Innovations in product design and manufacturing processes contribute to long-term competitiveness while minimizing the environmental footprint.

Investment in employee training programs to build awareness and understanding of sustainable practices, ensuring that the workforce is equipped to contribute to the organization's ESG goals.

Well-informed and engaged employees play a crucial role in implementing sustainability initiatives across various departments.

2. Human Capital – Nurturing the Seeds of Change

Establishment of specialized sustainability teams with expertise in environmental, social, and governance aspects to drive the implementation and monitoring of the ESG strategy.

A dedicated team ensures that sustainability is prioritized and initiatives are effectively executed with a focus on continuous improvement.

Attraction and retention of skilled professionals with expertise in sustainability, environmental science, social impact, and ethical governance.

A workforce with diverse skills related to sustainability is crucial for the successful execution and adaptation of the ESG strategy.

Engagement of external consultants and advisors with specialized knowledge in sustainability to provide guidance, conduct audits, and offer strategic insights.

External expertise ensures a fresh perspective and access to best practices, enhancing the effectiveness of sustainability initiatives.

3. Technological Infrastructure- Building the Foundation for Change

Implementation of technologies for real-time monitoring of environmental impacts, such as emissions tracking, water usage monitoring, and waste management systems.

Accurate data collection is essential for informed decision-making and meeting environmental targets.

Adoption of data analytics and reporting tools to measure and communicate key performance indicators, facilitating transparency and stakeholder engagement.

Efficient data management and reporting support the organization in tracking progress and demonstrating the impact of sustainability initiatives.

Integration of sustainable sourcing and supply chain management systems to trace and optimize the environmental and social impact of the entire value chain.

A transparent and sustainable supply chain is essential for achieving ESG goals and meeting stakeholder expectations.

4. Stakeholder Engagement- Cultivating Partnerships for Impact

Initiatives that involve financial support and active participation in community development programs ensure a positive social impact.

Building strong relationships with local communities contributes to a positive corporate reputation and enhances the organization's social license to operate.

Engagement with non-governmental organizations (NGOs) and industry groups to collaborate on initiatives, share best practices, and contribute to broader systemic change.

Partnerships amplify the impact of sustainability efforts and provide opportunities for collective action on shared challenges.

Establishment of regular communication channels, such as stakeholder forums, to actively engage with customers, employees, investors, and local communities.

Open communication fosters trust, ensures alignment with stakeholder expectations, and allows for valuable feedback that can inform strategy adjustments.

The Must-Win Battles

To successfully achieve sustainability, it is essential to identify and prioritize the "must-win battles." This process can be complex and requires careful navigation. Strategic priorities are crucial for achieving transformative and lasting impact. These battles serve as focal points, guiding organizations towards their overarching sustainability goals. Organizations can chart a course for lasting impact by addressing these priorities and contributing to a more sustainable, equitable, and resilient future.

1. Carbon Neutrality & Climate Action:

- **Objective:** Achieving carbon neutrality and actively mitigating climate change impacts.
- **Rationale:** Addressing climate change is a global imperative. Organizations must prioritize initiatives to reduce greenhouse gas emissions, transition to renewable energy sources, and contribute to global efforts to limit rising temperatures.

2. Circular Economy Adoption:

- **Objective:** Transitioning towards a circular economy by minimizing waste and maximizing resource efficiency.
- **Rationale:** Embracing circular practices, such as recycling, reusing, and reducing, is essential for minimizing environmental impact. This battle involves redesigning products, optimizing supply chains, and fostering a culture of sustainability.

3. Social Equity & Inclusion:

- **Objective:** Promoting diversity, equity, and inclusion across the organization and within the broader community.
- **Rationale:** A commitment to social equity contributes to a fair and inclusive workplace. Organizations must champion diversity in leadership, ensure fair labor practices, and actively engage with communities to address social inequalities.

4. Responsible Supply Chain Management:

- **Objective:** Ensuring responsible sourcing and supply chain practices, including ethical labor standards and sustainable sourcing of raw materials.
- **Rationale:** A transparent and ethical supply chain is integral to a sustainable business model. Organizations must assess and improve supply chain practices to align with environmental and social goals.

5. Biodiversity Conservation:

- **Objective:** Actively contributing to the conservation of biodiversity and ecosystems.
- **Rationale:** Biodiversity loss poses significant risks to ecosystems and human well-being. Organizations must engage in initiatives that protect and restore biodiversity, considering the impact of their operations on ecosystems.

6. Ethical Governance & Anti-Corruption:

- **Objective:** Upholding high standards of ethical governance and combating corruption.
- **Rationale:** Ethical governance builds trust with stakeholders and ensures the responsible operation of the organization. Implementing anti-corruption measures is crucial for maintaining integrity and ethical conduct.

7. Sustainable Product Innovation:

- **Objective:** Driving innovation for sustainable product design and development.
- **Rationale:** Developing products with a reduced environmental footprint and incorporating sustainable materials is key. This battle involves continuous research and development to align products with evolving sustainability standards.

8. Community Engagement & Impact:

- **Objective:** Actively engaging with and positively impacting local communities.
- **Rationale:** Building strong relationships with communities fosters a positive corporate reputation. Organizations must prioritize initiatives that contribute to community well-being and address local needs.

9. Employee Well-Being & Development:

- **Objective:** Ensuring the well-being, development, and satisfaction of employees.
- **Rationale:** Engaged and satisfied employees are essential for the successful implementation of sustainability initiatives. Organizations must prioritize employee well-being, provide development opportunities, and foster a positive workplace culture.

10. Transparent Communication & Reporting:

- **Objective:** Communicating sustainability efforts transparently to internal and external stakeholders.
- **Rationale:** Transparent communication builds trust and credibility. Organizations must proactively share their sustainability journey, progress, and challenges with stakeholders, demonstrating a commitment to accountability.

Clarifying dependencies

When it comes to sustainability, understanding and navigating dependencies is crucial for success. By clarifying dependencies, organizations can foster synergy, address challenges effectively, and create a resilient foundation for their sustainability strategy.

1. **Interdependence of Environmental, Social, and Governance (ESG) Components:**

- **Dependency:** The success of ESG strategies relies on recognizing the interdependence of environmental, social, and governance factors.
- **Clarification:** Environmental initiatives may impact social aspects, and ethical governance practices are integral to both environmental and social success. A comprehensive approach that considers these dependencies ensures a balanced and cohesive sustainability strategy.

2. Alignment With Regulatory Frameworks:

- **Dependency:** Adherence to sustainability goals often depends on alignment with regional and global regulatory frameworks.
- **Clarification:** Organizations must stay informed about evolving regulations related to environmental practices, social responsibility, and corporate governance. Aligning strategies with regulatory requirements ensures compliance and mitigates risks associated with non-compliance.

3. Integration With Corporate Culture:

- **Dependency:** The successful execution of sustainability initiatives depends on their alignment with the organization's culture and values.
- **Clarification:** Organizations must foster a culture that embraces sustainability at its core. Employee engagement, commitment to ethical practices, and a shared understanding of sustainability contribute to the seamless integration of sustainability into the organizational DNA.

4. Technological Infrastructure & Data Accuracy:

- **Dependency:** Accurate data is crucial for measuring and reporting on sustainability performance. The success of technology-driven initiatives depends on reliable data.
- **Clarification:** Implementing technologies for environmental monitoring, data analytics, and supply chain management relies on the availability of accurate and comprehensive data. Organizations must

invest in robust data management systems to ensure the reliability of sustainability metrics.

-

5. Stakeholder Engagement & Communication:

- **Dependency:** The success of sustainability efforts depends on effective stakeholder engagement and transparent communication.
- **Clarification:** Engaging with stakeholders, including employees, customers, investors, and local communities, is crucial for garnering support and feedback. Transparent communication builds trust and ensures that stakeholders understand and align with the organization's sustainability journey.

6. Financial Commitments & Resource Allocation:

- **Dependency:** Achieving sustainability goals requires financial commitments and the allocation of resources.
- **Clarification:** Organizations must prioritize sustainability in budgeting and resource allocation. Clear financial commitments demonstrate leadership's dedication to sustainability and provide the necessary resources for the implementation of initiatives.

Sustainability is a complex web of interdependent threads. Identifying these connections is key to understanding its patterns. It enables organizations to navigate challenges, leverage synergies, and build a resilient foundation for their sustainability strategy.

L'Oréal's ESG (Environmental, Social, and Governance) strategy, as exemplified by its "L'Oréal for the Future" program, serves as an inspiring reference for companies looking to align their operations with sustainability goals.[lxi] This program, initiated in 2020, underscores L'Oréal's commitment to being a proactive part of the solution to global challenges.

One key aspect of L'Oréal's strategy is transforming its business activities to reduce environmental impact. They set bold targets for 2030 related to climate change, water management, biodiversity preservation, and resource conservation, aligning with scientific recommendations and planetary boundaries. This approach goes beyond internal operations to include indirect impacts, such as supplier activities and consumer usage of products, showcasing a comprehensive sustainability mindset.

For instance, L'Oréal has significantly reduced CO2 emissions from its operated sites and increased renewable energy usage. They also prioritize inclusion and community support, as evidenced by their initiatives that have helped thousands of disadvantaged individuals find employment. Furthermore, their commitment to sustainable product development, as seen through the SPOT tool, ensures that new products prioritize environmental and social considerations.

L'Oréal's dedication to sustainability is further highlighted by its impressive achievements, such as consistently high rankings in CDP assessments and the continuous improvement of product sustainability profiles.

Overall, L'Oréal's holistic approach to ESG, encompassing environmental stewardship, social impact, and governance transparency, serves as a valuable example for companies seeking to integrate sustainability into their core business strategies and make meaningful contributions to a more sustainable world.

KONE's commitment to sustainability is evident in its daily operations and strategic focus areas. The company aims to enhance urban living by facilitating smooth and safe movement within and between buildings while prioritizing sustainable practices. Sustainability is not just an objective but a core value that guides KONE's actions and relationships with customers.[lxii]

In terms of environmental responsibility, KONE strives to be a leader in sustainability within and beyond its industry. The company recognizes the importance of reducing environmental impact and promoting eco-friendly solutions in its offerings and operations.

Beyond environmental concerns, KONE's sustainability focus areas encompass safety, quality, diversity, equity, inclusion, ethics, and compliance. These pillars reflect the company's holistic approach to sustainability, emphasizing the well-being of people and the planet and ethical business practices.

KONE's vision is centered around creating the best People Flow® experience, leveraging its 110-year expertise to improve urban life and contribute to smarter, more sustainable cities. By integrating sustainability into its core values and business strategy, KONE aims to be a trusted partner for customers throughout the building life cycle, supporting them in creating better living environments for communities worldwide.

Unilever's Climate Policy Engagement Review is a bold initiative calling upon industry associations to enhance their climate advocacy efforts significantly.[lxiii] As Unilever emphasizes the urgent need for collaborative action to combat climate change, the review reveals critical gaps and opportunities for improvement among the industry groups Unilever collaborates.

The review assessed 27 industry associations, highlighting that eight of them lacked any substantial engagement with climate policy, while four exhibited low engagement levels. Additionally, eight associations were found to be misaligned with Unilever's priority policy areas related to climate action and the Paris Agreement targets.

Unilever stresses the importance of industry associations translating their Paris Agreement alignment into concrete actions rather than mere rhetoric. Recognizing that collective action is essential to driving meaningful change, Unilever urges industry associations to step up their efforts in shaping climate policies that are scientifically aligned with limiting global warming to 1.5°C.

To address these gaps, Unilever outlines specific actions in the review, including revisiting climate policy positions, establishing climate subcommittees within associations, and increasing transparency around lobbying activities. Unilever emphasizes that associations must evolve from passive entities to proactive catalysts for positive policy change.

Moreover, Unilever's commitment to transparency extends to its relationships with industry associations, with plans to update the review annually to track progress. By holding associations accountable and advocating for robust climate policies, Unilever aims not only to achieve its own climate goals but also to inspire broader positive impacts across the business community.

ESG Strategy Is a Transformation Strategy

In today's dynamic and rapidly evolving business landscape, ESG considerations have emerged as crucial pillars for sustainable success. Companies worldwide are increasingly recognizing the importance of integrating ESG factors into their strategic decision-making processes.

ESG strategy is not merely about ticking boxes on a checklist; it represents a fundamental transformation in how organizations operate, how they engage with stakeholders, and how they create long-term value. At its core, ESG strategy is about transformation. It goes beyond traditional business practices and challenges organizations to rethink their approach to risk management, stakeholder engagement, and long-term sustainability. ESG factors encompass a wide range of issues, from climate change and resource scarcity to diversity and inclusion, human rights, and ethical business practices. By addressing these factors proactively, companies can mitigate risks, seize opportunities, and build resilience in an increasingly complex and interconnected world.

Implementing an ESG strategy effectively requires a robust change management process. Change management is key to successfully navigating the organizational shifts necessary to embed ESG considerations into the fabric of the business. This involves aligning leadership, engaging employees at all levels, and fostering a culture of continuous improvement and innovation. Change management helps organizations overcome resistance to change, build buy-in for ESG initiatives, and ensure that the strategy is integrated into everyday decision-making processes.

Moreover, ESG strategy is not just about changing policies or procedures; it is about transforming mindsets and instilling a culture of sustainability and responsibility throughout the organization. This requires a top-down commitment to ESG principles as well as bottom-up engagement from employees who are essential in driving change from within.

By embedding ESG values into the organization's DNA, companies can create a shared sense of purpose, inspire innovation, and foster collaboration across functions and departments. Leveraging ESG as a source of strategic advantage is a powerful way for companies to differentiate themselves in the market and drive long-term value creation. Organizations that embrace ESG principles can enhance their brand reputation, attract and retain top talent, and build stronger relationships with customers, investors, and other stakeholders. By demonstrating a commitment to sustainability, transparency, and ethical business practices, companies can gain a competitive edge and position themselves for long-term success in a rapidly changing world.

ESG strategy is more than a set of guidelines or best practices; it is a transformational journey that requires organizations to rethink their purpose, values, and impact on society and the environment. By embracing ESG as a strategic imperative, companies can drive innovation, enhance resilience, and create sustainable value for all their stakeholders. Change management plays a crucial role in successfully implementing the ESG strategy, as it helps organizations navigate the complexities of transformation, build a culture of sustainability, and leverage ESG as a source of competitive advantage in the marketplace.

Conclusion

Developing a comprehensive sustainability strategy is a crucial step towards building a more responsible and resilient organization. This strategy should be deeply rooted in the results of your assessments and aligned with your sustainability goals, which themselves are informed by the "As-is" analysis, market analysis, and materiality assessment.

The strategy should define clear ambitions, both in the long and medium term, with a focus on addressing the most material environmental, social, and governance (ESG) impacts of your operations. Prioritize initiatives within each pillar of ESG, ensuring that the chosen activities have clear objectives, targets, and Key Performance Indicators (KPIs) that can be tracked over time.

The link between an organization's ESG strategy and its broader corporate strategy is essential for driving meaningful and impactful sustainability initiatives. By integrating ESG considerations into the corporate strategy, organizations ensure that sustainability goals are not treated as separate objectives but are instead embedded in the core business objectives. This alignment creates a unified approach to organizational success, where sustainability becomes a strategic driver of long-term resilience and prosperity.

For example, if a corporate strategy emphasizes innovation, the ESG strategy can align by prioritizing initiatives that promote sustainable product design, eco-friendly packaging, and circular economy practices. By integrating ESG considerations into decision-making at all levels, particularly during product development, organizations can assess environmental impacts, evaluate social implications in the supply chain, and ensure ethical governance in sourcing.

Once the ESG strategy is developed, the company can test how well aligned the strategy is with the below statements to ensure it covers foundational aspects:

1. The ESG strategy is well integrated into the company's overall corporate vision and strategy.
2. ESG is or will be implemented in the company's decision-making, processes, and systems throughout the organization.
3. Technology and data are supporting the ESG strategy and journey.

4. There is buy-in from leadership—both strategic and operational—for the ESG strategy.
5. An internal network has been established or identified to support the implementation of the ESG strategy.

Transparent communication about how the ESG strategy contributes to the broader corporate strategy fosters stakeholder understanding and support. Through annual reports, corporate communications, and stakeholder engagements, organizations can articulate how achieving sustainability targets directly contributes to the overall resilience, reputation, and financial success of the business.

Continuous improvement is key to maintaining the effectiveness of sustainability initiatives. Establishing feedback loops, conducting regular stakeholder surveys, and staying informed about emerging trends ensure that the ESG strategy remains responsive to evolving challenges and opportunities. An adaptive approach allows organizations to incorporate new regulations, technological advancements, and stakeholder expectations while maintaining alignment with the corporate strategy.

To achieve sustainability goals, strategic investments and resource allocations are essential. Financial investments in renewable energy infrastructure, R&D initiatives for sustainable product development, and employee training programs build the foundation for long-term success. Human capital investments involve establishing specialized sustainability teams, attracting skilled professionals, and engaging external consultants to provide expertise. Technological infrastructure investments support real-time monitoring of environmental impacts, data analytics for reporting, and sustainable supply chain management.

Cultivating partnerships and engaging stakeholders are also crucial components of an effective ESG strategy. By collaborating with NGOs, industry groups, and local communities, organizations can amplify the impact of their sustainability efforts and contribute to broader systemic change. Must-win battles, such as carbon neutrality, circular economy adoption, and social equity promotion, serve as focal points for guiding organizations towards their overarching sustainability goals. Transparent communication and reporting ensure accountability and build trust with stakeholders, reinforcing the organization's commitment to sustainability.

In summary, a well-developed sustainability strategy is a road map that guides your organization towards a more sustainable future. It is a dynamic document that should evolve over time, responding to changing priorities, market conditions, and stakeholder expectations. By establishing a clear strategy, your organization can create lasting value for itself, its stakeholders, and the planet.

Chapter 6
Integration with Business Operations

"Integrating sustainability into every facet of business operations is a mandate for future success. When sustainability becomes synonymous with efficiency, innovation, and responsibility, businesses thrive while contributing positively to the world around them."

The path towards sustainability is not just about words; it requires a significant transformation in how organizations operate and interact with the world. To truly integrate sustainability into a company's DNA, it must be embedded in every aspect of its operations.

Sustainability begins with the supply chain, where an organization's decisions have an impact on the entire lifecycle of a product or service. From sourcing raw materials to distribution, companies must scrutinize each link in the supply chain for its environmental, social, and governance impact. This involves cultivating responsible relationships with suppliers, reducing carbon footprints, and ensuring that ethical labor practices are followed throughout the supply network.

In the field of sustainability, product design is a powerful tool for change. By designing products that have a minimal environmental impact, using eco-friendly materials, and prioritizing recyclability, companies can contribute to a circular economy.

Sustainable manufacturing is not just a trendy term; it is a strategic necessity. Companies should optimize their production processes to minimize waste, energy consumption, and environmental impact.

Engaging customers in the sustainability journey is not just about marketing; it is about empowering them to make informed, sustainable choices. Companies should educate and inspire their customers to make conscious consumption decisions. From transparent labelling to creating a culture of conscious consumption, businesses can play a critical role in shaping a consumer landscape that values sustainability.

Creating a Formal ESG Policy Is a Manifesto for Sustainable Commitment

Establishing a formal Environmental, Social, and Governance (ESG) policy is a crucial step for organizations committed to sustainability. This policy serves as a guiding manifesto, outlining the company's dedication to environmental stewardship, social responsibility, and governance excellence. It not only communicates the organization's values but also sets the tone for integrating sustainability into its core operations.

Key Components of a Formal ESG Policy

An ESG policy outlines a company's commitment to sustainability and responsible business practices. Some key components that should be included in an ESG policy are:

1. **Commitment to Sustainability:**

   - Clearly state the company's commitment to sustainable practices and responsible business conduct.

2. **Environmental Commitments**

   - **Climate Change:** Address how the company is managing its carbon footprint and contributing to efforts to combat climate change.
   - **Resource Management:** Articulate strategies for optimizing the use of natural resources, including water, energy, and raw materials. This could involve recycling initiatives, sustainable sourcing, and waste reduction measures.
   - **Waste Management:** Include strategies for reducing, reusing, and recycling waste to minimize environmental impact.
   - **Eco-Friendly Practices:** Outline the organization's commitment to adopting eco-friendly practices in its operations. This may

encompass sustainable product design, packaging, and manufacturing processes.

3. Social Responsibility

- **Fair Labor Practices:** Clearly communicate the organization's commitment to fair and ethical labor practices. This involves ensuring safe working conditions, fair wages, and adherence to international labor standards.
- **Diversity & Inclusion:** Define the organization's stance on diversity and inclusion, emphasizing the importance of creating a workplace that values and respects individuals of all backgrounds, ethnicities, genders, and abilities.
- **Employee Well-being:** Outline initiatives and policies that prioritize the well-being of employees, including health and safety programs, mental health support, and work-life balance considerations.
- **Community Engagement:** Articulate the organization's commitment to positively impacting the communities in which it operates. This could involve philanthropy, volunteering programs, and initiatives that address local social challenges.
- **Supply Chain:** Address how the company ensures ethical practices throughout its supply chain, including labor standards and human rights.

4. Governance Standards

- **Transparency:** Emphasize the importance of transparency in organizational practices. This includes clear communication with stakeholders about the company's ESG performance, goals, and challenges.
- **Accountability:** Establish mechanisms for accountability at all levels of the organization. This may involve regular assessments, audits, and reporting to ensure that ESG commitments are upheld.

- **Ethical Conduct:** Define ethical principles that guide decision-making processes within the organization. This includes avoiding corruption, bribery, and unethical business practices.
- **Legal Compliance:** Commit to complying with all relevant laws and regulations related to environmental protection, labor practices, and corporate governance.
- **Risk Management:** Address how the company identifies and manages ESG risks.

Unilever, consumer goods giant, is renowned for its Sustainable Living Plan. This comprehensive initiative outlines ambitious environmental goals, including reducing its environmental footprint and enhancing social impact. Unilever's commitment to sustainability is embedded in its business strategy, influencing everything from sourcing raw materials to product design.[lxiv]

Assessing & Managing ESG Risks & Opportunities In The Supply Chain

The supply chain is a critical arena for ESG considerations, where risks and opportunities abound. Assessing and managing ESG factors within the supply chain requires collaboration with suppliers, fostering a collective commitment to sustainability. This approach not only mitigates risks but also catalyzes positive change throughout the entire supply network.

- **Risk Assessment:**

o *Identifying Risks:* Conduct a thorough assessment to identify potential ESG risks within the supply chain, considering factors like environmental impact, labor practices, and ethical governance.

o *Mapping Opportunities:* Simultaneously, identify opportunities for improvement and positive ESG contributions within the supply chain.

- **Engaging Suppliers:**

o *Setting Standards:* Clearly communicate the organization's ESG standards to suppliers, emphasizing the importance of aligning with these principles.

o *Collaborative Improvement:* Work closely with suppliers to improve their ESG performance. This may involve providing resources, guidance, and incentives for adopting sustainable practices.

Nike, a global leader in athletic footwear and apparel, has undertaken initiatives to enhance sustainability in its supply chain. The company engages with suppliers to reduce environmental impact, improve labor conditions, and innovate sustainable materials. By setting clear standards and collaborating with suppliers, Nike has made strides in mitigating ESG risks while fostering positive change.[lxv]

Developing Sustainable Products Is a Path To Responsible Growth

The development of sustainable products and offerings is not only a moral imperative but also a strategic move that aligns business growth with environmental and social responsibility. Organizations should integrate sustainability into their product development strategies, creating value for both the business and the planet.

Sustainable Design Principles

- **Life Cycle Assessment:** Incorporate life cycle assessments into the design process to evaluate the environmental impact of products from raw material extraction to end-of-life disposal.
- **Eco-friendly Materials:** Prioritize the use of eco-friendly and renewable materials in product design. This may involve exploring alternatives to traditional materials that have a lower environmental footprint.
- **Circular Economy Practices:** Design products with circular economy principles, considering recyclability, reusability, and minimizing waste throughout the product's life cycle.

Innovation For Sustainability

- **Green Technologies:** Leverage innovative technologies that contribute to sustainability goals. This may include adopting clean energy sources, implementing energy-efficient processes, and incorporating smart technologies for resource optimization.

- **Biodegradable Solutions:** Explore biodegradable alternatives for products, packaging, and materials to minimize environmental impact and contribute to a circular economy.
- **Sustainable Supply Chains:** Collaborate with suppliers who adhere to sustainable practices, ensuring that the entire supply chain is aligned with environmental and social responsibility.

Customer Education & Engagement

- **Transparent Labelling:** Clearly communicate the sustainability features of products through transparent labelling. This helps customers make informed choices and reinforces the organization's commitment to transparency.
- **Consumer Awareness Campaigns:** Launch campaigns to educate customers about the environmental and social benefits of sustainable products. Increase awareness of the positive impact their choices can have on the planet.

The development of sustainable products and offerings is increasingly recognized as a strategic imperative for responsible growth, aligning business objectives with environmental and social responsibility. Organizations across various sectors are integrating sustainability into their product development strategies to create value for both their businesses and the planet.

For example, in the baking sector, companies like **King Arthur Baking Company** have embraced sustainability by offering organic and responsibly sourced baking ingredients. By prioritizing eco-friendly materials and transparent labelling, they cater to environmentally conscious consumers while reducing their environmental footprint. [lxvi,lxvii]

Similarly, **IKEA**, a global leader in furniture retail, has implemented sustainable design principles throughout its product range. Through initiatives like the "IKEA Sustainability Strategy," the company has committed to using renewable and recycled materials, reducing waste, and promoting circular economy practices.[lxviii,lxix] For instance, IKEA's "Better Cotton Initiative"[lxx,lxxi,lxxii] ensures the responsible sourcing of cotton for its products, contributing to environmental conservation and fair labor practices in the supply chain.

These examples demonstrate how organizations can leverage sustainable design principles, innovation, and customer engagement to develop products that align with environmental and social sustainability goals. By integrating sustainability into their product offerings, companies not only meet the growing demand for eco-friendly solutions but also contribute to a more sustainable future for generations to come.

Financial Impact of ESG Priorities

Understanding the financial impact of ESG priorities is crucial for organizations seeking to balance sustainability with profitability. ESG considerations affect the bottom line and provide insights into measuring and managing these financial implications.

1. Cost of Implementation:

Upfront Investments: Acknowledge the initial costs associated with implementing sustainable practices, such as upgrading technologies, obtaining eco-friendly certifications, and reconfiguring supply chains.

Long-term Savings: Consider the long-term cost savings resulting from sustainable initiatives, such as reduced energy consumption, lower waste disposal costs, and enhanced operational efficiency.

Implementing robust ESG practices helps companies steer clear of fines and penalties associated with non-compliance with environmental and governance regulations. For instance, businesses in China can face fines for contributing to environmental pollution or ecological damage. By adhering to ESG standards, companies can mitigate these financial risks. Investing in ESG initiatives often leads to operational efficiencies and cost savings. For example, reducing carbon emissions or optimizing water usage not only aligns with sustainability goals but also saves on utility costs in the long term. Similarly, improving health and safety protocols reduces injury-related costs and minimizes employee absences, contributing to overall operational efficiency.[lxxiii]

2. Revenue Opportunities:

Market Differentiation: Recognize the potential for market differentiation and increased customer loyalty that comes with being a sustainability leader. Consumers increasingly prefer brands that align with their values, creating revenue opportunities for organizations committed to ESG principles.

Innovation-Driven Growth: Embrace sustainability as a driver of innovation, leading to the creation of unique products and services that cater to evolving market demands.

Investors increasingly favor companies with strong ESG performance. By demonstrating commitment and progress in ESG areas, businesses can attract investment and access funding for sustainable growth initiatives. This interest extends beyond public companies to private investment as well. Companies with robust ESG credentials are more attractive to customers, especially in key markets like the US and Europe. Large companies seek business partners and suppliers aligned with their ESG goals, offering preferential terms to those with strong sustainability credentials. This presents revenue opportunities through enhanced customer relationships and partnerships. Some international buying companies incentivize their business partners to improve ESG performance by offering finance programs linked to sustainability metrics. This creates revenue opportunities for suppliers who align with ESG standards and demonstrate continuous improvement.[lxxiii]

According to a recent report from McKinsey, growing demand for net-zero offerings could generate more than USD 12 trillion of annual sales by 2030 across 11 value pools, including transport (USD 2.3 trillion to USD 2.7 trillion per year), power (USD 1.0 trillion to USD 1.5 trillion), and hydrogen (USD 650 billion to USD 850 billion). Such a transformation of the global economy could create significant growth potential for businesses through technologies, products, and services.[lxxiv]

3. Risk Mitigation:

Supply Chain Resilience: Consider the financial benefits of mitigating risks within the supply chain, such as avoiding disruptions due to climate-related events, regulatory changes, or reputational damage associated with unsustainable practices.

Legal & Regulatory Compliance: Factor in the potential financial risks of non-compliance with evolving environmental and social regulations. Proactively aligning with regulatory expectations can prevent costly legal challenges.

4. Stakeholder Relations:

Investor Confidence: Recognize the impact of ESG performance on investor confidence. Many investors now consider ESG factors when making investment decisions, and strong ESG performance can attract sustainable investment opportunities.

Employee Productivity & Retention: Acknowledge the financial implications of a positive corporate culture fueled by sustainability initiatives. Employee productivity, satisfaction, and retention can contribute to long-term financial stability.

Embedding ESG Considerations Into Day-To-Day Decision-Making

The integration of Environmental, Social, and Governance (ESG) considerations into the fabric of an organization represents a paradigm shift. It's a commitment to weave sustainability into the very DNA of the business, transcending the notion of ESG as a standalone initiative. Organizations should embed ESG considerations into day-to-day decision-making processes, business practices, and risk management, fostering a holistic and sustainable business strategy.

1. Integrating ESG Into Decision-Making:

Incorporating ESG considerations into decision-making processes is essential for organizations aiming to operate sustainably and responsibly. Strategic alignment ensures that ESG factors are seamlessly integrated into broader business objectives rather than treated as separate entities. Moreover, inclusive stakeholder input, encompassing perspectives from employees, customers, investors, and communities, enriches decision-making by providing a comprehensive understanding of ESG implications.

2. Business Practices Aligned With ESG:

Business practices must reflect a commitment to ESG principles throughout the organization's operations. This includes extending ESG considerations across the supply chain and collaborating with like-minded suppliers to promote responsible practices from sourcing to delivery. Additionally, fostering employee engagement through training and incentives ensures that sustainability becomes ingrained in daily work practices, reinforcing the organization's commitment to ESG.

3. Risk Management Through an ESG Lens:

An ESG-focused approach to risk management is crucial for identifying and addressing potential vulnerabilities. By incorporating ESG factors into the risk management framework, organizations can proactively identify environmental, social, and governance risks. Developing mitigation strategies that address both immediate concerns and long-term impacts enhances resilience and prepares the organization to navigate ESG challenges effectively.

4. ESG as a Core Business Strategy:

ESG considerations should not merely be peripheral to business strategy but rather integral to its core. This involves integrating ESG goals into strategic planning processes and setting specific, measurable, achievable, relevant, and time-bound (SMART) targets aligned with the company's long-term vision. Establishing key performance indicators (KPIs) to measure and track ESG performance allows organizations to assess progress regularly and adjust strategies to stay on course towards sustainability goals.

Unilever has been a pioneer in sustainability and has demonstrated a strong commitment to ESG principles across its operations. It has integrated ESG goals into its strategic planning processes. The company has set ambitious targets related to reducing environmental impact, promoting social responsibility, and enhancing governance practices. For example, they have committed to achieving net-zero emissions across their operations by 2039 and ensuring that all their plastic packaging is reusable, recyclable, or compostable by 2025.

Unilever sets specific, measurable, achievable, relevant, and time-bound (SMART) targets for each ESG focus area. These targets are aligned with the company's long-term vision of sustainable growth. For instance, they aim to source 100% of their agricultural raw materials sustainably and improve the livelihoods of millions of people in their value chain by 2025.

Unilever tracks its ESG performance using key performance indicators (KPIs). They measure metrics such as carbon emissions, water usage, waste generation, diversity and inclusion in the workforce, ethical sourcing practices, and community engagement initiatives.

Unilever conducts regular assessments of its ESG performance against KPIs. They analyze data, identify areas for improvement, and adjust strategies accordingly. For example, if they find that water consumption in a manufacturing facility exceeds targets, they implement water-saving technologies and initiatives to meet their sustainability goals.

5. Holistic Reporting & Transparency:

Finally, holistic reporting and transparent communication are essential for demonstrating accountability and building trust with stakeholders. Integrating ESG disclosures into corporate reports provides stakeholders with a comprehensive view of the organization's performance in terms of environmental impact, social responsibility, and governance. Transparent communication fosters an open dialogue, reinforcing the organization's commitment to ESG principles and accountability.

Unilever's Sustainable Living Plan exemplifies the integration of ESG considerations into the core of its business strategy. Unilever sets ambitious goals related to environmental impact reduction, social responsibility, and governance. By embedding these objectives into day-to-day decision-making and business practices, Unilever showcases how ESG can be a driving force for sustainable growth.[lxxv]

Nike is another company that has taken several initiatives to enhance sustainability in its supply chain. The company's Move to Zero initiative aims to achieve net-zero carbon emissions and zero waste by 2050. Nike is focusing on carbon, waste, water, and chemistry, aiming to hit targets by 2025. The company is increasing its use of environmentally preferred materials to 50% of all key materials, including polyester, cotton, leather, and rubber, to reduce greenhouse gas emissions by 0.5 million metric tons. Nike is also working to divert 100% of waste from landfills in its extended supply chain, with at least 80% of waste recycled back into Nike products and other goods. Additionally, the company is aiming to reduce fresh water usage per kilogram in textile dyeing and finishing by 25%.[lxxvi]

Empowering Through Education & Fostering An ESG-Centric Culture

Offering Environmental, Social, and Governance (ESG) training to employees transcends mere compliance; it's a pivotal strategy for nurturing a culture of responsibility and accountability within the organization. Rather than viewing ESG training as a checkbox exercise, organizations should recognize it as a means to instill a sustainable mindset among their workforce. Through education, businesses can empower employees to contribute to ESG initiatives actively, aligning their actions with broader organizational goals.

Crafting tailored ESG training programs is essential to ensure employees grasp the relevance of ESG principles to their roles and the organization's mission. These programs should encompass a comprehensive curriculum covering environmental impact, social responsibility, and governance practices. Employing interactive learning methods, such as workshops and webinars, fosters engagement and enables employees to apply ESG principles effectively in their work.

Employee involvement and engagement are paramount in fostering a culture of sustainability. Encouraging employees to participate in decision-making processes related to ESG initiatives promotes a sense of ownership and commitment. Facilitating cross-functional collaboration breaks down silos and fosters a shared understanding of sustainability goals across departments.

Linking ESG principles to personal impact reinforces individual accountability and commitment to sustainability. Employees should understand how their daily responsibilities contribute to the organization's ESG objectives. Recognizing and rewarding employees for their contributions to ESG initiatives reinforces the value placed on sustainable practices within the organization.

Leadership buy-in and role modelling are instrumental in driving ESG initiatives forward. Visible support from top leadership, coupled with their active participation in ESG training programs, demonstrates organizational commitment to sustainability. Leaders should embody ESG principles in their practices, serving as role models for employees to emulate.

Continuous learning and updates ensure that employees stay abreast of evolving ESG trends and best practices. Providing regular updates on emerging ESG topics and soliciting employee feedback fosters a culture of continuous improvement and ensures that ESG programs remain relevant and effective over time.

Google has prioritized sustainability education and engagement among its employees. The company offers various training programs, workshops, and resources on environmental sustainability, encouraging employees to incorporate sustainable practices into their work. Google's commitment to ESG is not just a top-down directive; it's a shared responsibility embraced by every employee.[lxxvii lxxviii]

Sustainability Is a Call To Infuse in Every Aspect of Your Business

Embracing sustainability is not merely a choice; it represents a transformative journey towards a future where every aspect of business contributes to a thriving planet and society. It's a commitment that transcends profit margins, calling for a fundamental shift in how we approach products, processes, and practices.

Consider the profound impact of eliminating single-use plastics from your operations and products. By taking this pledge, you not only contribute to a healthier environment but also set a precedent for responsible business practices. Embracing sustainable packaging alternatives heralds a future where ecological consciousness is woven into every product.

Introducing electric vehicles into your fleet is another significant step towards a greener horizon. Beyond being an eco-friendly choice, it signifies a commitment to reducing carbon footprints and embracing sustainable transportation solutions. By embracing electric vehicles, you showcase a dedication to innovative, environmentally conscious practices.

Nurturing a sustainable culture within your organization is essential for long-term success. Imagine a workplace where every employee is not just a contributor but a champion of sustainability. By enhancing employee engagement through comprehensive sustainability training, you empower your workforce to integrate environmental and social responsibility into their daily roles. This fosters a collective force for positive change and ensures that sustainability becomes ingrained in the organizational ethos.

Moreover, consider the societal benefits that extend beyond profit margins. Your business has the potential to become a catalyst for positive societal change, whether through philanthropy, community outreach programs, or support for local initiatives. By actively contributing to the well-being of the communities you serve, you elevate your purpose and create a legacy that extends far beyond financial success.

Why Make This Commitment?

1. **Future-Proof Your Business:** Sustainability isn't just a trend; it's the future of business. By embedding sustainable practices now, you future-proof your organization, staying ahead of evolving consumer expectations and regulatory requirements.
2. **Attract & Retain Talent:** In a world where employees seek purpose in their work, a commitment to sustainability becomes a powerful magnet for talent. It's not just about what your business does; it's about the positive impact it creates in the world.
3. **Build Trust With Consumers:** Consumers today are discerning, seeking products and services aligned with their values. Embedding sustainability builds trust, creating a loyal customer base that values businesses contributing to a better world.
4. **Contribute To Global Goals:** By integrating sustainability, your business becomes a contributor to global sustainability goals, making a tangible impact on issues like climate change, plastic pollution, and social inequality. (More details in chapter 5)

The examples of companies like **Eastman Kodak Company, Polaroid Corporation, Blockbuster Inc., and Borders Group** serve as stark reminders of what can happen to businesses that fail to innovate and adapt to market changes. These companies were once market leaders but faced bankruptcy due to their inability to embrace disruptive innovation and evolving consumer preferences. Their stories offer valuable lessons, especially in the context of ESG considerations, which are becoming increasingly crucial in today's business landscape.[lxxix]

Eastman Kodak Company, for instance, was renowned for its cameras and photographic film, popularizing the concept of a "Kodak moment." However, despite being a pioneer in imaging technology, Kodak failed to foresee the digital revolution and adapt its business model accordingly. The rise of digital cameras led to a decline in demand for traditional film-based products, causing financial turmoil and eventual bankruptcy for Kodak.

Similarly, **Polaroid Corporation**, known for its instant photography, faced a similar fate with the advent of digital photography. The company struggled to innovate and lost its market relevance as consumers shifted towards digital cameras and photo-sharing platforms.

Blockbuster Inc., a giant in the video rental industry, failed to anticipate the digital streaming era and the popularity of online entertainment platforms like Netflix. Blockbuster's reluctance to embrace digital distribution and adapt its business model ultimately led to its downfall and bankruptcy.

Borders Group, a major player in the bookstore business, faced challenges from e-commerce and digital reading devices like the Kindle. The company failed to pivot towards online sales and e-books, unlike its competitor, Barnes & Noble, and eventually succumbed to bankruptcy.

These examples highlight the dire consequences of ignoring market changes and failing to innovate. Just as these companies faced disruptions in their industries, businesses that neglect ESG considerations risk falling behind in today's rapidly evolving business environment. Embracing ESG principles is not just about ethical responsibility but also about staying relevant, attracting investors, winning customer trust, and mitigating risks associated with regulatory compliance and changing consumer preferences.

In short, the fate of companies that did not innovate serves as a cautionary tale for businesses today. Embracing ESG as a core part of business strategy is not just a moral imperative but also a strategic necessity to thrive in a dynamic and competitive market landscape.

Resources allocated to ESG

Allocating and investing in ESG initiatives internally is crucial for companies looking to integrate sustainability into their core business strategies. One effective approach involves establishing standalone ESG resources dedicated to setting holistic direction and maintaining an overview of the organization's sustainability efforts. These resources are supported by multiple internal units to ensure comprehensive implementation and alignment with company goals.

1. **Establishment of Standalone ESG Resources**:

   o **ESG Department or Team**: Create a dedicated ESG department or team comprising professionals with expertise in sustainability, environmental management, social responsibility, and corporate governance. This team serves as the central hub for all ESG-related activities within the organization.

   o **Chief Sustainability Officer or Director**: Appoint a Sustainability/ESG officer or director who reports directly to senior management or the board of directors. This individual oversees the ESG strategy, initiatives, and performance, ensuring alignment with corporate objectives and external standards.

2. **Roles and Responsibilities**:

   o **Developing ESG Strategy**: The ESG team collaborates with various departments to develop a comprehensive ESG strategy aligned with the company's mission, values, and long-term goals.

   o **Setting Performance Metrics**: Define key performance indicators (KPIs) and metrics to measure the company's ESG performance across environmental, social, and governance domains.

   o **Monitoring and Reporting**: Regularly monitor ESG initiatives, collect data, and prepare detailed reports for internal stakeholders, investors, regulators, and external rating agencies.

   o **Engaging with Stakeholders**: Engage with stakeholders, including investors, employees, customers, suppliers, and communities, to communicate ESG efforts, gather feedback, and address concerns.

3. **Collaboration with Internal Units**:

   o **Finance and Investment Teams**: Collaborate with finance and investment teams to integrate ESG criteria into investment decisions, capital allocation, and risk management processes. This includes evaluating ESG risks and opportunities associated with investments.

- o **Human Resources**: Work with HR departments to implement ESG-focused training programs, diversity and inclusion initiatives, employee well-being programs, and ethical conduct guidelines.
- o **Operations and Supply Chain**: Partner with operations and supply chain teams to promote sustainable practices, reduce environmental impact, enhance supply chain transparency, and ensure responsible sourcing.
- o **Marketing and Communications**: Coordinate with marketing and communications teams to communicate ESG achievements, sustainability goals, and corporate responsibility efforts to internal and external audiences.

4. **Integration and Alignment**:

- o **Integration with Corporate Governance**: Align ESG policies, practices, and disclosures with corporate governance frameworks, regulatory requirements, and industry standards.
- o **Continuous Improvement**: Continuously assess and improve ESG performance through feedback mechanisms, benchmarking against peers, and staying updated on emerging ESG trends and best practices.
- o **Board Oversight**: Provide regular updates and reports to the board of directors or relevant committees on ESG performance, risks, opportunities, and strategic initiatives.

By establishing standalone ESG resources supported by multiple internal units, companies can effectively drive sustainable practices, enhance stakeholder value, mitigate risks, and contribute positively to society and the environment while achieving long-term business success.

Emergence of CSO[lxxx]

The emergence of the Chief Sustainability Officer (CSO) role represents a significant shift in corporate governance and strategy, marking a departure from the historical perception of CSOs as primarily focused on public relations and risk mitigation. The evolving role of CSOs is now centered around integrating material ESG issues into corporate strategy, aligning with broader business goals, and engaging directly with stakeholders and investors.

One of the primary drivers behind this transformation is the growing recognition among investors and executives that sustainability plays a crucial role in financial performance. Investors are increasingly factoring ESG criteria into their decision-making processes, prompting companies to rethink their approach to sustainability. Additionally, a changing political landscape, with debates around "woke capitalism" and corporate responses to global challenges has further emphasized the importance of sustainability as a strategic imperative rather than a peripheral concern.

The new CSO role involves collaborating closely with senior leadership teams, including the CEO, CFO, and investor relations, to integrate ESG considerations into strategic planning and capital allocation. CSOs are now more involved in investor meetings, reflecting a deeper engagement with investors who seek clarity on how sustainability efforts contribute to long-term value creation.

Furthermore, CSOs are advocating for four major changes to their role: involvement in strategy and capital allocation, realistic stakeholder interactions, deeper engagement with investors, and sufficient resources and expertise throughout the organization. This shift underscores the need for CSOs to move beyond traditional CSR initiatives and focus on sustainable value creation that aligns with business objectives.

The CSO's role in stakeholder engagement has also evolved, emphasizing strategic interactions with stakeholders critical to the company's business model. This targeted engagement approach allows companies to address key stakeholder concerns effectively while prioritizing areas that drive innovation and value creation.

Moreover, the CSO's collaboration with governance, risk, ethics, and compliance functions has become essential for ensuring alignment between sustainability initiatives and core business operations. This integration helps companies navigate complex ethical and governance challenges while maintaining a focus on sustainable growth.

To support these changes, CSOs require high-caliber talent with diverse backgrounds, including expertise in sustainability, business strategy, finance, and technology. Collaborative efforts across business units and functions are essential for embedding sustainability throughout the organization and driving long-lasting value.

Your Journey Towards a Lasting Impact

As you explore the possibilities of embedding sustainability into every facet of your business, envision the legacy you're creating—a legacy of responsible practices, positive societal impact, and a planet that thrives alongside your success. It's not just a choice; it's a commitment to a future where every business decision reflects the values of sustainability and leaves a lasting imprint on the world. Your journey towards lasting impact begins with a single, purposeful step—are you ready to take it?

Conclusion

The integration of sustainability into business operations marks a pivotal shift towards responsible corporate citizenship. This integration aligns with global trends and growing stakeholder expectations. A formal ESG policy signals your company's commitment to establishing clear principles that guide decision-making.

Moreover, embedding sustainability principles into core products and services amplifies this commitment. Companies can prioritize the development of sustainable offerings that not only meet consumer demand for eco-friendly solutions but also contribute to positive environmental and social impacts. For instance, a clothing manufacturer may implement sustainable practices throughout its supply chain, from sourcing organic materials to minimizing waste in production processes. Similarly, a technology company could design energy-efficient products that reduce carbon emissions and promote responsible consumption.

By integrating sustainability into core products and services, organizations not only enhance their reputation as responsible corporate citizens but also contribute to broader efforts aimed at addressing pressing environmental and social challenges. This strategic approach not only meets market demands but also fosters long-term sustainability and resilience in an increasingly interconnected global economy.

Managing ESG risks and opportunities in the supply chain demonstrates a commitment to ethical sourcing and responsible business practices. This proactive approach benefits not only your company but also suppliers and their communities. Developing sustainable products aligns with changing consumer preferences and presents opportunities for growth and innovation.

Understanding the financial impact of ESG priorities allows for strategic planning, ensuring that sustainable practices contribute to long-term profitability. By embedding ESG considerations into daily operations and decision-making, sustainability becomes inherent in your organization's DNA.

Providing employee training and encouraging their involvement fosters a culture of responsibility and accountability, empowering employees to contribute to sustainability initiatives. Actively seeking ways to embed sustainability throughout products and processes, such as reducing single-use plastics and embracing electric vehicles, showcases a commitment to creating benefits for both the environment and society at large.

Chapter 7
Measurement & Reporting

"Measuring and reporting on sustainability is about accountability and continuous improvement. Numbers tell a story, and when it comes to sustainability, that story shapes our future."

Establishing a robust system for measuring and tracking progress against sustainability goals is paramount for organizations committed to driving meaningful change. This process involves setting clear and measurable key performance indicators (KPIs) that align with the organization's sustainability objectives. These KPIs serve as benchmarks, providing tangible metrics to evaluate progress over time. For instance, KPIs may include metrics related to energy consumption, waste reduction, carbon emissions, diversity and inclusion, and community impact. By defining KPIs that reflect the organization's sustainability priorities, stakeholders gain clarity on what success looks like and can track performance effectively.

Collecting data on KPIs requires a systematic approach to data management and analysis. Organizations must establish processes for gathering relevant data from various sources, such as internal operations, supply chain partners, and community stakeholders. This data collection process may involve implementing tracking systems, conducting regular audits, and leveraging technology solutions for data aggregation and analysis. Additionally, organizations should ensure data accuracy and reliability by adhering to standardized measurement protocols and validation procedures. Organizations can generate accurate insights into their sustainability performance by centralizing data collection efforts and investing in data quality assurance measures.

Reporting on sustainability performance is a critical component of transparency and accountability to internal and external stakeholders.

Organizations should develop comprehensive reporting frameworks that communicate progress, challenges, and achievements related to sustainability goals. Internal reporting mechanisms allow for regular review and evaluation of sustainability initiatives by key decision-makers within the organization. External reporting, such as sustainability reports, annual disclosures, and stakeholder engagement platforms, provides investors, customers, regulators, and the broader community transparency. Through transparent reporting, organizations demonstrate their commitment to responsible business practices and build trust with stakeholders, ultimately driving continuous improvement in sustainability performance.

Determining Material ESG Factors

Determining the Environmental, Social, and Governance (ESG) factors that are material to your business and stakeholders is a critical step in developing a comprehensive sustainability strategy. Materiality assessment involves identifying the issues that have significant impacts on the organization's performance, reputation, and long-term viability, as well as those that matter most to stakeholders. Common ESG factors include carbon emissions, diversity metrics, employee turnover rates, community engagement, and governance practices.

For example, a technology company may identify carbon emissions as a material ESG factor due to the environmental impact of its data centers and operations. By measuring and tracking carbon emissions, the company can set targets to reduce its carbon footprint, invest in renewable energy sources, and improve energy efficiency.

Similarly, diversity metrics may be material to a financial institution, where workforce diversity and inclusion contribute to innovation, talent retention, and customer satisfaction. By monitoring diversity KPIs such as gender and ethnic diversity in leadership positions, employee turnover rates, and representation in recruitment, the organization can identify areas for improvement and implement initiatives to foster a more inclusive workplace culture.

Community Engagement & Governance Practices

Community engagement is another material ESG factor for industries with significant local or global impact, such as mining, manufacturing, or retail. Companies may measure community engagement through metrics such as philanthropic contributions, stakeholder consultations, or social impact assessments. Organizations can build trust and mitigate the social risks associated with their operations by actively engaging with local communities, addressing their concerns, and supporting sustainable development initiatives.

Governance practices, including board diversity, executive compensation, and transparency in decision-making, are essential for maintaining investor confidence and regulatory compliance. For example, a pharmaceutical company may establish governance KPIs related to board independence, ethical business conduct, and compliance with industry regulations. By demonstrating strong governance practices, companies can enhance their reputation, attract investment, and mitigate legal and reputational risks.

Establishing Specific & Measurable Targets

Setting specific and measurable targets for each identified Key Performance Indicator (KPI) is essential for effectively tracking progress towards sustainability goals. Specific targets provide clarity and direction, outlining exactly what needs to be achieved, while measurable targets allow for a quantifiable performance assessment. These targets should be aligned with the organization's overall sustainability objectives and take into account the materiality of each ESG factor.

For example, a manufacturing company may set a specific target to reduce water consumption by 20% within the next three years. This target is measurable as it quantifies the reduction in water usage, allowing the company to track its progress over time. Similarly, a retail company may establish a specific target to increase the representation of women in leadership positions to 30% within five years. This target is measurable through regularly monitoring the percentage of women in senior management roles.

Benchmarking Against Industry Peers & Best Practices

Benchmarking performance against industry peers and best practices provides valuable context for evaluating progress and identifying areas for improvement. By comparing performance metrics with similar organizations within the industry, companies can assess their relative standing and identify opportunities to excel or address gaps. Benchmarking against best practices allows organizations to learn from leaders in sustainability and adopt strategies and initiatives that have proven effective elsewhere.

For instance, an automotive manufacturer may benchmark its carbon emissions per vehicle produced against industry peers to gauge its environmental performance. By comparing its emissions intensity with that of top-performing competitors, the company can identify opportunities to reduce its carbon footprint and improve operational efficiency. Similarly, a financial institution may benchmark its diversity metrics, such as the percentage of women in leadership roles, against industry leaders to identify areas for improvement and set ambitious yet achievable targets.

Survey Of Sustainability Reporting

KPMG's latest survey examines sustainability reporting trends globally. They analyzed financial, sustainability, and ESG reports for the top 100 companies across 58 countries, territories, and jurisdictions.

Technology, media, and telecommunications (TMT) companies constitute 10% of this group.

TMT companies have made significant progress in reporting carbon reduction targets. 81% of TMT companies now disclose climate and carbon reduction targets, up from 61% in 2017.[lxxxi]

According to the KPMG CEO Outlook, 55% of technology and telecommunication CEOs believe that ESG programs enhance their company's financial performance. This represents a substantial increase from last year (38% for tech companies and 28% for telcos).

Sustainability initiatives and reporting are now long-term commitments that leaders should allocate human and financial resources toward.[lxxxi]

Big Tech's Net-Zero Goals

Companies like Amazon, Google, and Microsoft have ambitious net-zero carbon emissions goals.

They are investing in carbon accounting, climate risk analysis, and other climate tech areas.

These giants are committed to sustainability and transparent reporting.[lxxxii]

Another example comes from the consumer goods industry, where companies like Unilever and Procter & Gamble benchmark their sustainability performance against best practices. For instance, Unilever's Sustainable Living Plan sets ambitious targets for reducing environmental impact and improving social conditions, benchmarking progress against leading sustainability frameworks such as the Sustainable Development Goals (SDGs) and the Science-Based Targets initiative.

In conclusion, establishing specific and measurable targets for each identified KPI and benchmarking performance against industry peers and best practices are essential steps in tracking progress towards sustainability goals. By setting ambitious yet achievable targets and learning from industry leaders, organizations can drive continuous improvement and make meaningful contributions to environmental, social, and governance objectives.

Gathering Relevant Data for ESG Performance Tracking

Tracking Environmental, Social, and Governance (ESG) performance requires gathering relevant data from both internal and external sources. Internal data collection involves gathering information directly from the organization's operations and processes, such as energy usage, waste generation, and employee demographics. External data encompasses industry standards, regulatory requirements, and other external benchmarks against which the organization's performance can be evaluated. (More details in chapter 2.)

Internal Data Collection

The internal data collection process is a critical step in any organization's sustainability journey. By collecting and analyzing internal data, companies can gain insights into their environmental impact, resource efficiency, and workforce diversity. There are three key areas of internal data collection, and each of these areas plays a crucial role in shaping a company's sustainability strategy. Energy Usage data helps identify opportunities for energy efficiency improvements; Waste Generation data provides insights into waste reduction initiatives; and Employee Demographics data informs diversity and inclusion strategies. Through the collection and analysis of internal data, companies can develop targeted and impactful sustainability initiatives that align with their organizational goals and values.

Energy Usage: Monitoring energy consumption is crucial for assessing environmental impact and identifying opportunities for improvement. This includes tracking electricity, gas, and water usage across different facilities and operations. Organizations may use smart meters, utility bills, or energy management systems to collect data on energy consumption.

Waste Generation: Tracking waste generation and disposal provides insights into resource efficiency and environmental sustainability. Organizations can collect data on the types and quantities of waste generated, as well as recycling and waste diversion rates. This data helps identify opportunities for waste reduction and recycling initiatives.

Employee Demographics: Understanding the demographic composition of the workforce is essential for assessing diversity and inclusion efforts. Data on employee demographics, including gender, ethnicity, age, and job roles, can help identify disparities and inform diversity and inclusion strategies.

External Data Collection

The external data collection process involves gathering information from external sources to supplement internal data and provide a comprehensive view of an organization's sustainability performance. There are two key areas of external data collection. Benchmarking ESG performance against industry standards allows organizations to assess their performance relative to peers while ensuring compliance with regulatory requirements, which is critical for legal and regulatory compliance. By collecting and analyzing external data, organizations can gain valuable insights into their sustainability performance, identify areas for improvement, and ensure compliance with industry standards and regulations.

Industry Standards: Benchmarking ESG performance against industry standards allows organizations to assess their performance relative to peers and identify areas for improvement. Industry-specific metrics and benchmarks provide context for evaluating performance and setting targets. Organizations may use industry associations, sustainability frameworks, and industry reports to gather relevant data.

Regulatory Requirements: Compliance with regulatory requirements is critical for ensuring legal and regulatory compliance and managing operational risks. Organizations must track relevant environmental, social, and governance regulations and ensure that their practices align with regulatory standards. This may involve collecting data on emissions, hazardous materials, labor practices, and governance practices to ensure compliance with applicable laws and regulations.

Walmart, a global retail giant, is committed to sustainability and transparent reporting. Here's how they collect and assess data for their **Environmental, Social, and Governance (ESG)** performance:

- **ESG Reporting & Priorities**:

Walmart's **ESG priorities** are organized into four leadership themes: **Opportunity**, **Sustainability**, **Community**, and **Ethics & Integrity**.

They focus on areas such as **human capital**, **climate change**, **waste reduction**, and **animal welfare**.[lxxxiii]

Walmart's **ESG reports** provide detailed insights into their progress and initiatives.

- **Data Collection**:

Walmart gathers extensive data on various aspects:

> - **Energy Usage**: They track energy consumption across their operations.
> - **Waste Generation**: Walmart monitors waste generated and implements circular economy practices.
> - **Supply Chain Sustainability**: They collect data from suppliers to assess environmental impact.

Internal data from operations and suppliers is crucial for evaluating performance.

- **External Benchmarks & Standards**:

Walmart compares its performance against **industry benchmarks** and **global reporting frameworks**.

They align their ESG reporting with the **Global Reporting Initiative (GRI) Standard** and other relevant frameworks.[lxxxiv]

- **Specific Achievements**:

Walmart has achieved:

> - A **23.2% annual reduction** in Scopes 1 and 2 emissions since 2015.
> - Donations of **665 million pounds of food** to fight hunger.
> - Engagements with shareholders representing **550 million shares**.
> - Safe driving of over **1 billion miles** by Walmart fleet drivers.
> - Sourcing from **diverse suppliers** and recyclable packaging initiatives.[lxxxiii]

In summary, Walmart's data-driven approach, alignment with global standards, and commitment to sustainability contribute to its positive ESG performance.

In conclusion, gathering relevant data is essential for tracking ESG performance and identifying opportunities for improvement. By collecting internal data on energy usage, waste generation, and employee demographics and leveraging external benchmarks and regulatory requirements, organizations can assess their performance, set targets, and drive continuous improvement in environmental, social, and governance practices.

Assessing ESG Performance Against KPIs & Targets

Once data on environmental, social, and governance (ESG) performance has been collected, organizations must analyze this information to assess their progress against established key performance indicators (KPIs) and targets. This assessment involves evaluating performance metrics, identifying trends, and pinpointing areas for improvement.

Data Analysis & Evaluation

Data analysis and evaluation are crucial steps in the sustainability assessment process, enabling organizations to gauge their performance against KPIs and identify trends that signal progress or areas of concern. This phase is instrumental in pinpointing specific areas that need improvement, allowing organizations to allocate resources effectively and drive meaningful change. Overall, it plays a pivotal role in shaping sustainability strategies and fostering continuous improvement.

Performance Metrics: Organizations analyze collected data to evaluate performance metrics such as carbon emissions, energy efficiency, diversity metrics, community engagement, and governance practices. By comparing actual performance against predefined KPIs and targets, organizations can determine whether they are on track to achieve their sustainability objectives.

Identifying Trends: Data analysis helps organizations identify trends in ESG performance over time. By examining historical data and tracking changes in performance metrics, organizations can identify patterns and trends that indicate progress or areas of concern. For example, decreasing energy consumption may indicate successful energy efficiency initiatives, while increasing employee turnover rates may signal challenges in workforce management.

Areas For Improvement: Data analysis also reveals areas where performance may fall short of targets or industry benchmarks. By identifying gaps between current performance and desired outcomes, organizations can prioritize areas for improvement and develop targeted strategies and action plans to address these deficiencies. For instance, if data analysis reveals high carbon emissions relative to industry peers, the organization may implement measures to reduce its carbon footprint, such as investing in renewable energy sources or improving energy efficiency.

Involving Stakeholders in the Measurement Process

Involving internal and external stakeholders in the measurement process is essential for gaining diverse perspectives and insights into the social and environmental impact of the organization. Stakeholder engagement fosters transparency, accountability, and collaboration, enhancing the organization's overall sustainability efforts.

Internal Stakeholder Involvement

Employees: Employees play a crucial role in driving sustainable practices within the organization. Involving employees in the measurement process empowers them to contribute ideas, provide feedback, and participate in initiatives aimed at improving ESG performance. Employee perspectives can offer valuable insights into workplace practices, culture, and engagement levels, helping identify opportunities for improvement.

External Stakeholder Engagement

Investors: Investors increasingly consider ESG factors when making investment decisions. Engaging with investors on ESG performance metrics and targets demonstrates the organization's commitment to sustainability and transparency. Investors may provide feedback and insights that inform strategic decisions and help attract investment capital aligned with ESG principles.

Customers: Customers are increasingly interested in supporting businesses that demonstrate social and environmental responsibility. Involving customers in the measurement process through surveys, feedback mechanisms, and product transparency initiatives allows organizations to understand customer preferences and priorities related to sustainability. Customer feedback can inform product development, marketing strategies, and brand reputation management.

Community Members: Engaging with local communities and other external stakeholders provides valuable perspectives on the social and environmental impact of the organization's operations. Community input can inform decisions related to corporate social responsibility initiatives, community engagement programs, and sustainable business practices. Organizations demonstrate their commitment to responsible business practices and build trust with stakeholders by involving community members in the measurement process.

Assessing ESG performance against established KPIs and targets requires comprehensive data analysis to identify trends and areas for improvement. Involving internal and external stakeholders in the measurement process fosters transparency, accountability, and collaboration, enhancing the organization's ability to address social and environmental challenges and achieve sustainable outcomes.

Developing a Comprehensive ESG Report

Creating an Environmental, Social, and Governance (ESG) report is pivotal in demonstrating an organization's commitment to transparency and accountability regarding its sustainability performance. A comprehensive ESG report provides stakeholders with valuable insights into the organization's ESG practices, impacts, and progress towards sustainability goals. Here's how organizations can develop an effective ESG report:

Identify Key ESG Factors: Begin by identifying the most material ESG factors relevant to your organization and stakeholders. These may include environmental metrics such as carbon emissions and water usage, social indicators like employee diversity and community engagement, and governance practices such as board diversity and ethical business conduct. Prioritize factors that have the most significant impact on your organization's operations and stakeholders.

Data Collection & Analysis: Gather relevant data on identified ESG factors from internal sources, such as operational records and employee surveys, as well as external sources, such as industry benchmarks and regulatory reports. Analyze this data to assess your organization's performance against key metrics and targets. Identify strengths, weaknesses, and areas for improvement and ensure data accuracy and reliability.

Transparency & Balance: Ensure that your ESG report is transparent, balanced, and accessible to stakeholders. Present information in a clear, concise, and understandable format, using charts, graphs, and narratives to illustrate key findings and trends. Provide context for the data presented, including explanations of methodologies, assumptions, and limitations. Acknowledge both successes and challenges and be honest about areas where improvement is needed.

Stakeholder Engagement: Engage with internal and external stakeholders throughout the ESG reporting process to gather input, address concerns, and ensure that the report reflects their interests and priorities. Seek feedback on draft versions of the report to incorporate diverse perspectives and improve its relevance and usefulness to stakeholders.

Integration With Financial Reporting: Integrate ESG reporting with financial reporting to provide a comprehensive view of the organization's overall performance and value creation. Highlight connections between ESG factors and financial outcomes, such as the impact of sustainability initiatives on cost savings, revenue growth, and risk mitigation. Align ESG reporting with established financial reporting standards and frameworks to ensure consistency and comparability.

Adhering to Globally Recognized ESG Reporting Frameworks

To enhance the credibility and comparability of ESG reporting, organizations should adhere to globally recognized ESG reporting frameworks and standards. These frameworks provide guidance on ESG disclosure, measurement, and reporting practices, helping organizations improve transparency, accountability, and stakeholder trust. Some widely recognized ESG reporting frameworks include:

Global Reporting Initiative (GRI): The GRI framework is one of the most widely used standards for ESG reporting globally. It provides a comprehensive set of guidelines and indicators for reporting on economic, environmental, and social performance, as well as governance practices. Adhering to GRI standards helps organizations disclose material ESG information in a structured and standardized manner, facilitating comparability and benchmarking.

Sustainability Accounting Standards Board (SASB): SASB standards focus on industry-specific ESG factors that are financially material to companies within specific sectors. By aligning ESG disclosure with industry-specific metrics, SASB standards help organizations tailor their reporting to the unique risks and opportunities they face. This enhances the relevance and usefulness of ESG information for investors, analysts, and other stakeholders.

Task Force on Climate-related Financial Disclosures (TCFD): The TCFD framework provides recommendations for disclosing climate-related financial risks and opportunities in mainstream financial filings. By integrating climate-related information into financial reporting, organizations can better inform investors and stakeholders about the potential impacts of climate change on their business operations and financial performance. TCFD-aligned reporting helps organizations assess and disclose climate-related risks and opportunities in a consistent and transparent manner, supporting more informed decision-making and risk management.

GHG Protocol: The GHG Protocol, developed by the World Resources Institute (WRI) and the World Business Council for Sustainable Development (WBCSD), provides globally recognized methodologies for quantifying and managing greenhouse gas emissions. It consists of three scopes, each defining a different emission source.

Scope 1 refers to direct emissions originating from sources under the control of the organization, such as emissions from fuel combustion in boilers or company-owned vehicles. Scope 2 covers indirect emissions generated from the consumption of purchased electricity, steam, heating, and cooling, while Scope 3 includes all other indirect emissions arising from activities such as business travel, waste disposal, and employee commuting, often known as value chain emissions. By considering these three scopes, organizations can understand the full spectrum of their emissions and implement strategies to mitigate their environmental impact.

International Sustainability Standards Board (ISSB): The ISSB is an initiative established by the International Financial Reporting Standards (IFRS) Foundation to develop globally recognized sustainability disclosure standards. These standards aim to enhance the consistency, comparability, and credibility of sustainability reporting, providing investors, regulators, and other stakeholders with consistent and reliable information on companies' sustainability performance. The ISSB is responsible for developing a set of global sustainability disclosure standards that are aligned with the objectives of the Paris Agreement and the Sustainable Development Goals (SDGs). These standards are expected to cover a wide range of sustainability topics, including environmental, social, and governance (ESG) factors. The ISSB standards will provide a common language for sustainability reporting, enabling companies to communicate their sustainability performance in a transparent and standardized manner. This will facilitate better decision-making by investors, enhance market efficiency, and ultimately contribute to a more sustainable and resilient global economy. By adhering to ISSB standards, organizations can ensure that their sustainability disclosures are aligned with global best practices, enhancing the credibility and comparability of their ESG reporting. This will enable stakeholders to make more informed decisions and drive positive change towards a more sustainable future.

Task Force on Nature-related Financial Disclosures (TNFD): The TNFD is a globally recognized ESG reporting framework focused on nature-related risks and opportunities. It provides guidance to organizations on disclosing, measuring, and reporting nature-related financial information, similar to how other frameworks address environmental, social, and governance (ESG) factors.

The TNFD framework aims to enhance transparency and accountability regarding nature-related impacts, such as biodiversity loss, deforestation, water security, and climate change's impact on ecosystems. By integrating nature-related considerations into financial reporting, organizations can better assess and manage risks associated with their operations, supply chains, and investments.

TNFD's framework encourages organizations to identify and assess nature-related risks and opportunities, develop strategies to mitigate risks and capitalize on opportunities, and disclose relevant information to stakeholders. This approach helps companies align their business activities with sustainable practices, foster resilience against environmental challenges, and build trust among investors, regulators, and the broader community.

Adhering to the TNFD framework enables organizations to demonstrate their commitment to environmental stewardship, drive positive environmental outcomes, and contribute to a more sustainable global economy. It also facilitates comparability and consistency in nature-related reporting, allowing stakeholders to make informed decisions and incentivizing responsible business practices.

Developing a comprehensive ESG report and adhering to globally recognized ESG reporting frameworks are essential steps for organizations committed to transparent and accountable sustainability practices. Organizations can enhance trust, credibility, and long-term value creation by providing stakeholders with clear, balanced, and standardized information on their ESG performance.

Integrating ESG Disclosures Into Financial Reporting

Integrating Environmental, Social, and Governance (ESG) disclosures into an organization's financial reporting processes is a strategic approach to highlighting the interconnectedness of financial and non-financial performance. By incorporating ESG information into annual reports, financial statements, and other financial disclosures, organizations can provide stakeholders with a more comprehensive view of their overall performance and value creation. Here's how organizations can effectively integrate ESG disclosures into their financial reporting:

Alignment With Financial Materiality: Identify ESG factors that are financially material to the organization and its stakeholders. Materiality assessments help determine which ESG issues have the greatest impact on financial performance, risk management, and long-term value creation. Focus on disclosing ESG information that is relevant, reliable, and decision-useful for investors, analysts, and other stakeholders.

Integration Into Annual Reports: Embed ESG disclosures within annual reports, which serve as primary communication channels for stakeholders interested in the organization's financial and non-financial performance. Incorporate ESG-related narratives, metrics, and key performance indicators (KPIs) into the Management Discussion and Analysis (MD&A) section, business overview, and strategic outlook to provide context and insights into ESG-related opportunities, risks, and impacts.

Enhanced Financial Statements: Supplement traditional financial statements with additional disclosures related to ESG performance. For example, include environmental-related disclosures in the notes to the financial statements, such as information on greenhouse gas emissions, energy consumption, and water usage. Highlight social and governance-related factors that may impact financial performance, such as workforce diversity, human rights practices, and corporate governance structures.

Consistency & Comparability: Ensure consistency and comparability of ESG disclosures with financial reporting standards and guidelines. Adhere to globally recognized ESG reporting frameworks, such as the Global Reporting Initiative (GRI) and the Sustainability Accounting Standards Board (SASB), to provide standardized and comparable information. Align ESG reporting with established financial reporting principles, such as relevance, reliability, and materiality, to enhance credibility and transparency.

Obtaining External Assurance for ESG Reporting

Obtaining external assurance for an organization's ESG report from a reputable third party can enhance the credibility, reliability, and trustworthiness of the reported ESG information. External assurance involves an independent assessment of the organization's ESG disclosures, processes, and controls by a qualified assurance provider. Here are key considerations for obtaining external assurance for ESG reporting:

Selection Of An Assurance Provider: Choose a reputable assurance provider with expertise in ESG reporting and assurance standards. Look for assurance providers that are accredited, independent, and experienced in conducting ESG assurance engagements. Consider factors such as industry expertise, geographical coverage, and a track record of delivering high-quality assurance services.

Scope Of Assurance Engagement: Define the scope and objectives of the assurance engagement, including the ESG disclosures and performance metrics to be assured, the reporting period covered, and the criteria against which the assurance will be conducted. Ensure that the scope is aligned with stakeholder expectations, regulatory requirements, and best practices in ESG reporting and assurance.

Assurance Methodology: Determine the assurance methodology and procedures to be followed, taking into account the nature, complexity, and materiality of the ESG disclosures. Select appropriate assurance standards, such as the International Standard on Assurance Engagements (ISAE) 3000 or the AA1000 Assurance Standard, and tailor the assurance approach to suit the organization's specific needs and objectives.

Reporting Of Assurance Findings: Communicate the findings of the assurance engagement transparently and objectively. Prepare an assurance report that clearly states the scope, objectives, and conclusions of the assurance engagement, along with any limitations or qualifications identified. Provide stakeholders with assurance statements or opinions that attest to the reliability and accuracy of the organization's ESG disclosures.

By integrating ESG disclosures into financial reporting processes and obtaining external assurance for ESG reporting, organizations can demonstrate their commitment to transparency, accountability, and stakeholder trust. These practices help investors, analysts, and other stakeholders make informed decisions, assess risks, and evaluate the long-term sustainability and resilience of the organization.

Tracking Progress Towards Goals & Strategy Implementation

Tracking progress towards sustainability goals and strategy implementation is essential for assessing the effectiveness of initiatives and ensuring alignment with organizational objectives.

Internally, organizations must create robust internal reporting mechanisms to ensure the effectiveness of sustainability initiatives and alignment with organizational objectives. This involves establishing clear key performance indicators (KPIs) and metrics to measure performance against targets. Regular monitoring and evaluation enable organizations to identify areas of success, challenges, and opportunities for improvement. By leveraging data analytics and reporting tools, organizations can track progress in real-time, enabling timely adjustments to strategies and actions to stay on course towards achieving sustainability objectives.

One such tool is the GHG Protocol, which provides a framework for quantifying and managing greenhouse gas emissions. This protocol allows organizations to identify and manage their environmental impact effectively, supporting their overall sustainability efforts. By measuring emissions across scopes, companies can understand their carbon footprint comprehensively and set targets for reduction or offsetting.

Another key framework is the International Sustainability Standards Board (ISSB). This board develops standards for sustainability-related financial reporting, providing guidance on how companies can integrate ESG factors into their financial disclosures. These standards help organizations communicate their sustainability performance transparently and effectively, enhancing stakeholder trust and accountability.

Moreover, organizations can establish robust internal audit processes to review and verify sustainability data, ensuring accuracy, reliability, and compliance with reporting standards. By adopting these measures, companies can enhance transparency, accountability, and credibility in their sustainability reporting, driving continuous improvement and progress towards their sustainability goals.

Regulatory Required Reporting on ESG & Stakeholder Expansion

Meeting regulatory requirements for ESG reporting is critical for compliance and transparency. Organizations must adhere to relevant regulations and standards, such as mandatory reporting frameworks or industry-specific guidelines. Additionally, expanding reporting beyond regulatory requirements to address stakeholder expectations is essential for building trust and credibility. Engaging stakeholders in the reporting process helps identify material ESG issues, enhance disclosure quality, and demonstrate responsiveness to stakeholder concerns. Organizations can enhance transparency and accountability by aligning reporting practices with stakeholder needs and preferences.

Transparency

Transparency is a cornerstone of effective ESG reporting and communication. Organizations must disclose relevant information openly and honestly to stakeholders, including investors, customers, employees, and communities. Transparency builds trust and credibility, fosters positive relationships, and enhances reputation. This involves providing clear and accurate data on ESG performance, highlighting successes, challenges, and areas for improvement. By being transparent about their sustainability practices, organizations can demonstrate integrity, accountability, and a commitment to responsible business conduct.

Third-Party Assurance

Obtaining third-party assurance for ESG reporting adds credibility and reliability to the disclosed information. Independent assurance providers assess the accuracy, completeness, and reliability of ESG disclosures, providing stakeholders with confidence in the reported data. Third-party assurance enhances transparency, accountability, and trust by validating organizational claims and assertions. It helps identify reporting processes and control gaps, enabling organizations to enhance disclosure quality and integrity. By seeking third-party assurance, organizations demonstrate a commitment to accountability and credibility in their sustainability reporting practices.

Establishing Internal ESG Reporting Similar to Financial Reporting

Establishing internal ESG reporting processes that mirror financial reporting practices enables organizations to effectively manage and communicate ESG performance internally. This involves integrating ESG data collection, analysis, and reporting into existing management systems and processes. By adopting standardized reporting frameworks and protocols, organizations can streamline data collection and ensure consistency in reporting practices. Internal ESG reporting facilitates decision-making, performance management, and risk mitigation by providing actionable insights into sustainability performance. By treating ESG reporting with the same rigor and discipline as financial reporting, organizations can enhance transparency, accountability, and stakeholder confidence in their sustainability efforts.

Ten Places to Focus in ESG Reporting

1. **Carbon Emissions:** Measure and report greenhouse gas emissions across scopes to assess environmental impact and track progress towards emission reduction targets.
2. **Energy Efficiency:** Monitor energy consumption and efficiency initiatives to identify opportunities for resource conservation and cost savings.
3. **Diversity & Inclusion:** Track workforce demographics, diversity metrics, and inclusion initiatives to promote a diverse and inclusive workplace culture.
4. **Supply Chain Management:** Assess supplier performance, ethical sourcing practices, and supply chain transparency to mitigate risks and enhance sustainability throughout the value chain.
5. **Community Engagement:** Report on community initiatives, philanthropic activities, and social impact projects to demonstrate

commitment to corporate social responsibility (CSR) and stakeholder engagement.

6. **Governance Practices:** Disclose governance structures, policies, and practices to ensure transparency, accountability, and ethical conduct in decision-making processes.

7. **Water Management:** Measure water usage, wastewater discharge, and water stewardship efforts to address water-related risks and promote responsible water management.

8. **Employee Well-Being:** Monitor employee health, safety, and well-being initiatives to create a safe and supportive work environment and foster employee engagement and satisfaction.

9. **Product Sustainability:** Assess product lifecycle impacts, eco-design strategies, and sustainable product innovations to minimize environmental footprints and meet customer expectations for sustainability.

10. **Regulatory Compliance:** Ensure compliance with relevant ESG regulations, reporting requirements, and industry standards to mitigate legal and reputational risks and maintain a license to operate.

By focusing on these key areas in ESG reporting and management, organizations can enhance their sustainability performance, build stakeholder trust, and create long-term value for all stakeholders.

Conclusion

In establishing a robust system for measuring and reporting on sustainability performance, organizations work towards greater transparency, accountability, and stakeholder engagement. By determining the material ESG factors relevant to their business and stakeholders, organizations lay the foundation for setting specific and measurable targets. Benchmarking against industry peers and best practices provides valuable context for evaluating progress and identifying areas for improvement. Gathering relevant data, both internally and externally, ensures a comprehensive assessment of ESG performance, encompassing factors such as carbon emissions, diversity metrics, and community engagement.

Engaging internal and external stakeholders in the measurement process enriches the evaluation with diverse perspectives and insights into the social and environmental impact of the organization. Developing a transparent and balanced ESG report, adhering to globally recognized reporting frameworks, and integrating ESG disclosures into financial reporting processes underscore the interconnectedness of financial and non-financial performance. Seeking external assurance further enhances the credibility and reliability of reported information, instilling stakeholder trust and confidence.

Organizations must prioritize regulatory compliance, transparency, and stakeholder engagement as they track progress towards their sustainability goals and strategy implementation. Embracing transparency and establishing internal ESG reporting mechanisms akin to financial reporting demonstrates a commitment to accountability and integrity. Organizations can drive meaningful change, create long-term value, and contribute to a more sustainable future for all by focusing on key areas such as carbon emissions, diversity, and community engagement.

Furthermore, with the emerging focus on nature and biodiversity in the field of sustainability, organizations are recognizing the importance of integrating nature-related considerations into their reporting frameworks. The Task Force on Nature-related Financial Disclosures (TNFD) provides valuable guidance in this regard. TNFD's framework encourages organizations to assess and disclose information related to their impacts and dependencies on nature and biodiversity. This includes factors such as biodiversity loss, deforestation, water security, and climate change's impact on ecosystems.

By incorporating nature-related metrics and disclosures into their ESG reporting, organizations can enhance their understanding of nature-related risks and opportunities. This enables them to develop more robust strategies for mitigating risks, promoting biodiversity conservation, and aligning their business activities with sustainable practices. Transparent reporting on nature-related issues also fosters trust among stakeholders, including investors, regulators, customers, and the broader community.

Chapter 8
Continuous Improvement

"Continuous improvement in sustainability is not a destination but a journey. Each step forward, no matter how small, contributes to a better world for future generations."

Implementing a process for continuous improvement is not just a strategic imperative but a cultural commitment embedded within the ethos of organizations striving for sustainability. It signifies an ongoing journey towards betterment, where learning from both successes and challenges is paramount. By embracing a dynamic approach to sustainability, organizations foster adaptability, resilience, and innovation in the face of evolving environmental, social, and economic landscapes.

Continuous improvement in sustainability involves a cycle of reflection, evaluation, and action. It starts with a willingness to acknowledge both achievements and shortcomings, leveraging successes as catalysts for further progress while treating challenges as opportunities for growth. By fostering a culture that values feedback and encourages experimentation, organizations create an environment conducive to learning and innovation.

Learning from successes involves identifying what worked well and understanding the underlying factors contributing to positive outcomes. Whether it's the successful implementation of energy efficiency measures, the launch of a community engagement initiative, or the achievement of diversity targets, organizations can extract valuable insights to replicate and scale effective practices. Moreover, celebrating successes fosters morale, reinforces organizational values, and inspires further commitment to sustainability goals.

Similarly, challenges present invaluable learning opportunities that shape the trajectory of sustainability strategies. Whether it's encountering obstacles in supply chain sustainability, facing regulatory hurdles, or navigating stakeholder resistance, organizations can glean insights into areas requiring improvement. By embracing a mindset of resilience and adaptability, organizations can pivot strategies, refine approaches, and overcome obstacles with agility and determination.

Staying Informed About the Latest ESG Trends

In today's rapidly evolving landscape, where societal expectations, market dynamics, and regulatory frameworks are constantly shifting, organizations must remain vigilant and adaptive. By continuously assessing their strategy in light of new information, organizations can ensure its relevance and effectiveness in driving sustainable outcomes.

One crucial aspect of staying informed is keeping abreast of emerging ESG trends and best practices. This involves monitoring industry publications, attending conferences, and participating in webinars or workshops focused on sustainability. By learning from leading experts and innovators in the field, organizations can gain valuable insights into emerging issues, innovative solutions, and evolving stakeholder expectations. Moreover, benchmarking against industry peers allows organizations to identify gaps in their strategy and adopt cutting-edge practices that drive continuous improvement.

Additionally, staying informed about regulatory developments is essential for maintaining compliance and anticipating future requirements. Governments and regulatory bodies worldwide are increasingly enacting laws and standards aimed at promoting sustainability, transparency, and responsible business practices. By closely monitoring regulatory changes and engaging with relevant authorities, organizations can proactively adjust their strategy to align with emerging requirements and mitigate compliance risks. Furthermore, engaging with industry networks and sustainability initiatives provides opportunities for collaboration, knowledge sharing, and collective action on pressing sustainability challenges.

In summary, staying informed about ESG trends, best practices, and regulatory developments is imperative for organizations committed to sustainability. By remaining proactive and adaptive, organizations can leverage new information to refine their strategy, enhance their performance, and drive positive impact across environmental, social, and governance dimensions. In a rapidly changing world, continuous learning and adjustment are essential for staying ahead of the curve and maximizing the effectiveness of sustainability efforts.

Establishing a Yearly Assessment Process

This process involves systematically reviewing and evaluating your organization's performance, initiatives, and goals over the past year to identify achievements, challenges, and areas for improvement. By conducting regular assessments, organizations can track their sustainability trajectory, measure the effectiveness of their efforts, and make informed decisions to steer their strategy in the right direction.

The yearly assessment begins by defining clear evaluation criteria aligned with your sustainability goals and key performance indicators (KPIs). These criteria may include environmental impact metrics, social responsibility initiatives, governance practices, stakeholder engagement outcomes, and progress towards long-term sustainability targets. By establishing specific benchmarks and metrics for evaluation, organizations can ensure consistency and comparability in their assessments over time. (More details in chapter 3.)

During the assessment process, gathering and analyzing relevant data and performance indicators to measure progress accurately is essential. This may involve collecting data on energy consumption, waste generation, carbon emissions, employee engagement, diversity metrics, community outreach activities, regulatory compliance, and other relevant factors. By leveraging data-driven insights, organizations can identify trends, patterns, and areas of strength or weakness in their sustainability performance.

Furthermore, the yearly assessment provides an opportunity for reflection and learning from past experiences. Organizations can conduct thorough reviews of their sustainability initiatives, projects, and strategies to identify successes, challenges, and lessons learned. By celebrating achievements and acknowledging areas for improvement, organizations can foster a culture of continuous learning and innovation. Additionally, engaging stakeholders, including employees, investors, customers, suppliers, and community members, in the assessment process can provide valuable perspectives and feedback on the organization's sustainability journey.

Based on the assessment's findings, organizations can develop action plans and recommendations for refining and enhancing their sustainability strategy in the coming year. This may involve setting new goals, revising existing targets, realigning priorities, allocating resources more effectively, and implementing corrective actions to address identified gaps or shortcomings. By incorporating insights from the assessment into strategic decision-making processes, organizations can drive continuous improvement and maximize the impact of their sustainability efforts year after year.

For example, consider a multinational retail corporation that conducts an annual sustainability assessment across its global operations. Upon analyzing the assessment results, the company identifies areas where it can further reduce its carbon footprint, such as optimizing logistics and supply chain operations. In response, it sets new emissions reduction targets for the upcoming year, invests in renewable energy sources for its facilities, and collaborates with suppliers to implement more sustainable transportation practices. Additionally, the company revises its product packaging to use more eco-friendly materials, aligning with consumer preferences for environmentally responsible products.

Similarly, a tech startup specializing in software development conducts a sustainability assessment to evaluate its environmental impact and social responsibility practices. Upon reviewing the assessment findings, the company recognizes the need to enhance its diversity and inclusion initiatives within the workplace. As a result, it develops a comprehensive action plan to recruit and retain a more diverse workforce, implement inclusive hiring practices, and provide training on unconscious bias and cultural sensitivity. By integrating these initiatives into its sustainability strategy, the company aims to foster a more inclusive work environment while driving positive social impact.

In both examples, the organizations use insights from their sustainability assessments to inform strategic decision-making and drive meaningful change. By proactively addressing sustainability challenges and seizing opportunities for improvement, these companies demonstrate their commitment to responsible business practices and long-term value creation.

In conclusion, establishing a yearly assessment of your sustainability journey is essential for maintaining momentum, accountability, and effectiveness in pursuing sustainability goals. By systematically evaluating performance, learning from experiences, and engaging stakeholders, organizations can refine their strategy, drive positive change, and contribute to a more sustainable future. The yearly assessment serves as a valuable tool for guiding decision-making, enhancing transparency, and demonstrating progress towards sustainability objectives over time.

Internal Reporting & Knowledge Sharing With Stakeholders

These processes ensure alignment, accountability, and engagement across all levels, fostering a culture of transparency, shared responsibility, and continuous improvement.

Reporting to leadership involves providing regular updates and insights on the organization's sustainability performance, progress towards goals, and key initiatives. This communication serves to keep leadership informed about the organization's sustainability journey, successes, challenges, and areas for improvement. By presenting data-driven metrics, case studies, and impact assessments, sustainability leaders can demonstrate the value and importance of sustainability efforts to the overall business strategy. Reports to leadership should be clear, concise, and tailored to the audience's interests and priorities, highlighting both the financial and non-financial benefits of sustainable practices. Moreover, it's essential to engage in dialogue with leadership, seeking their input, support, and guidance in driving sustainability initiatives forward and integrating sustainability considerations into decision-making processes.

Collaborating with stakeholders on knowledge sharing involves actively engaging with internal and external stakeholders to exchange information, insights, and best practices related to sustainability. This collaborative approach fosters a culture of learning, innovation, and collective action, leveraging the expertise and resources of diverse stakeholders. Internally, knowledge-sharing initiatives may include employee training sessions, workshops, webinars, and forums where employees can learn about sustainability trends, initiatives, and success stories. By empowering employees with knowledge and skills, organizations can build a more informed and motivated workforce that actively contributes to sustainability goals. Externally, organizations can engage with stakeholders such as customers, suppliers, investors, NGOs, and industry partners to share knowledge, collaborate on sustainability projects, and address shared challenges. Through partnerships, networks, and collaborative platforms, organizations can leverage collective intelligence and resources to drive systemic change and accelerate progress towards sustainability goals.

In conclusion, reporting to leadership and collaborating with stakeholders on knowledge sharing are essential strategies for advancing sustainability within organizations. Organizations can secure the necessary resources, alignment, and commitment to drive meaningful change by keeping leadership informed, engaged, and supportive of sustainability efforts. Similarly, by engaging stakeholders in knowledge-sharing initiatives, organizations can build capacity, foster collaboration, and leverage collective expertise to address complex sustainability challenges effectively. Together, these strategies enable organizations to embed sustainability into their DNA, achieve meaningful impact, and contribute to a more sustainable and prosperous future for all.

For instance, consider a multinational manufacturing company that implements robust internal reporting mechanisms to track its sustainability performance. Through regular updates to leadership, the company provides insights into its progress towards sustainability goals, highlighting achievements and areas for improvement. By presenting data-driven metrics and case studies, the company demonstrates the business value of sustainability initiatives, such as energy efficiency improvements or waste reduction strategies. Leadership receives clear, concise reports tailored to their interests, enabling informed decision-making and support for sustainability initiatives.

Similarly, a technology firm engages in knowledge-sharing initiatives with its stakeholders to enhance sustainability practices. Internally, the company conducts training sessions and workshops to educate employees about sustainability trends and best practices. Through collaborative forums and webinars, employees share insights and success stories, fostering a culture of innovation and continuous learning. Externally, the company collaborates with suppliers, customers, and industry partners to address sustainability challenges collectively. For example, the firm partners with suppliers to improve supply chain transparency or collaborate with customers to develop eco-friendly product solutions. These collaborative efforts leverage collective expertise and resources to drive positive change and accelerate progress towards sustainability goals.

Raising Internal & External Awareness

Internally, raising awareness involves educating employees about the importance of sustainability, the organization's sustainability goals and initiatives, and their role in contributing to these efforts. This can be achieved through various channels, such as training sessions, workshops, newsletters, intranet articles, and internal communication campaigns. By increasing awareness among employees, organizations can cultivate a sense of ownership and responsibility towards sustainability, encouraging them to adopt more sustainable behaviors both at work and in their personal lives. Moreover, raising awareness internally helps to build a shared understanding of sustainability concepts, values, and best practices across different departments and levels of the organization, fostering a culture of collaboration and innovation.

Unilever, a multinational consumer goods company, has implemented various strategies to educate and engage its employees in sustainability practices.[lxxxv]

Unilever conducts extensive training programs and workshops to educate employees at all levels about sustainability issues and the company's Sustainable Living Plan, which outlines ambitious environmental and social targets. These training sessions cover topics such as sustainable sourcing, waste reduction, water conservation, and the importance of biodiversity conservation.

Furthermore, Unilever utilizes internal communication channels, including newsletters, intranet platforms, and employee forums, to disseminate information about sustainability initiatives, progress reports, and success stories.[lxxxvi] The company also encourages open dialogue and feedback from employees, fostering a culture of transparency and collaboration.

In addition to formal training and communication channels, Unilever organizes various engagement activities and events to raise awareness about sustainability. For example, the company hosts sustainability-themed workshops, green challenges, and volunteer opportunities[lxxxvii] for employees to participate in environmental conservation projects.

Through these initiatives, Unilever ensures that its employees understand the importance of sustainability and are empowered to integrate sustainable practices into their daily work routines and personal lives. This internal awareness and engagement contribute to Unilever's broader sustainability strategy and its commitment to creating a more sustainable future for people and the planet.

Externally, raising awareness about sustainability involves engaging with stakeholders such as customers, suppliers, investors, regulators, and the broader community to communicate the organization's sustainability commitments, initiatives, and impact. This can be done through various communication channels, such as corporate websites, sustainability reports, social media, public events, and stakeholder engagement forums. By sharing information about their sustainability efforts, organizations can enhance their reputation, build trust with stakeholders, and attract like-minded partners and investors. Moreover, raising awareness externally helps to demonstrate organizational transparency, accountability, and leadership in addressing pressing sustainability challenges, contributing to broader awareness and action on sustainability issues in society.

One exemplary initiative aimed at raising external awareness about sustainability issues is the "Frontiers in Sustainability" program launched by **First Abu Dhabi Bank (FAB)** in collaboration with the Emirates Foundation and the International Institute for Management Development (IMD).[lxxxviii] This unique executive education initiative targets senior corporate leaders in the UAE, including both private and governmental entities, with the goal of promoting a deeper understanding of sustainability and its integration into business strategies. Through tailored executive learning sessions led by experts from world-renowned institutions like IMD, participants gain valuable insights and practical knowledge on sustainability best practices. The program's launch in 2023 coincides with the UAE's declaration of the year as the 'Year of Sustainability,' demonstrating a strategic alignment with the nation's broader sustainability goals. By offering strategic guidance and knowledge-sharing opportunities, FAB, along with its partners, aims to empower UAE corporates to embrace sustainable business practices, thus contributing to the country's transition towards a more sustainable and resilient future. Through initiatives like "Frontiers in Sustainability," FAB demonstrates its commitment to driving external awareness and fostering collaboration among stakeholders to address pressing sustainability challenges and achieve collective progress towards a net-zero economy by 2050.

Upskilling of Staff To Support the Transition

This involves providing training and development opportunities to equip employees with the knowledge, skills, and competencies needed to thrive in a rapidly changing business landscape characterized by environmental challenges and opportunities. Green skills encompass a wide range of competencies related to sustainability, including but not limited to energy efficiency, renewable energy technologies, circular economy principles, eco-design, sustainable supply chain management, climate resilience, and environmental management systems.

Organizations can offer training programs, certifications, workshops, and mentoring opportunities to help employees develop these green skills and integrate sustainability into their roles and responsibilities. By upskilling staff, organizations can enhance their capacity to innovate, adapt, and drive positive change towards a more sustainable future while also creating a more engaged and empowered workforce that is better equipped to navigate the complexities of the green economy.

An exemplary initiative aimed at upskilling staff to support the transition towards sustainability is **BBVA's** pioneering sustainability training program for its global workforce of over 125,000 employees.[lxxxix] As the first major bank in the world to mandate sustainability training for all employees, BBVA demonstrates a commitment to equipping its workforce with the necessary knowledge and skills to address pressing environmental and social challenges. The comprehensive training covers various aspects of sustainability, with a particular focus on climate change and its implications for both society and the finance industry. By providing employees with a strategic perspective on sustainable development and finance, BBVA ensures that its workforce understands the importance of integrating sustainability into the organization's operations and decision-making processes. The training program aims to raise awareness about global challenges, such as climate change, and empower employees to contribute to the achievement of Sustainable Development Goals (SDGs). Moreover, BBVA's training model offers flexibility and autonomy to employees, allowing them to access a diverse range of instructional resources, including online platforms, videos, seminars, and podcasts. By putting employees in the driver's seat of their own learning journey, BBVA fosters a culture of continuous learning and professional development, ultimately enabling its workforce to play an active role in advancing the bank's sustainability agenda. Through strategic partnerships and ongoing refinement of training materials, BBVA ensures that its training program remains responsive to the evolving sustainability knowledge needs of employees across different roles, functions, and geographic locations. By investing in the upskilling of its staff, BBVA reinforces its commitment to responsible business practices and positions itself as a leader in sustainable finance within the banking industry.

Conclusion

Maintaining a proactive approach to sustainability is paramount in today's dynamic business landscape. Staying informed about the latest ESG trends, best practices, and regulatory developments ensures that organizations remain adaptable and responsive to evolving expectations. By continuously assessing their sustainability strategy in light of new information, organizations can make necessary adjustments to enhance its relevance and effectiveness, driving long-term value creation and resilience.

Establishing a yearly assessment of the sustainability journey provides a structured framework for evaluating progress, identifying areas for improvement, and celebrating successes. Regular reporting to leadership keeps stakeholders informed and engaged, fostering accountability and alignment with organizational goals. Collaborating with stakeholders on knowledge sharing enhances focus and generates collective insights, driving meaningful action towards shared sustainability objectives.

Raising awareness about sustainability both internally and externally cultivates a culture of responsibility and innovation while ensuring that upskilling staff equips organizations with the capabilities needed to navigate the transition to a green economy. Together, these strategies empower organizations to embrace sustainability as a strategic imperative, driving positive impact and creating a more sustainable future for all stakeholders.

Chapter 9
Communication & Transparency

"Communication and transparency in sustainability efforts are the bridges that connect organizations with stakeholders, fostering trust, accountability, and shared progress toward a sustainable future."

One of the key aspects of being a sustainable organization is communicating your sustainability efforts and progress transparently to your stakeholders, both internally and externally. This can help build trust and support for your sustainability journey, which is essential for the long-term success of your sustainability initiatives.

Transparency is critical in today's world, where people are becoming more aware of the impact their actions have on the environment and society. By sharing information about your sustainability efforts, you can demonstrate your commitment to sustainability, build credibility, and foster a sense of trust among your stakeholders.

When communicating your sustainability efforts, it's important to be clear, concise, and consistent. You should use language that is easy to understand and avoid technical jargon that may confuse your audience. You should also provide relevant data and metrics that demonstrate your progress towards your sustainability goals.

In addition to communicating your sustainability efforts externally, it's also important to keep your internal stakeholders informed. This includes your employees, suppliers, shareholders, and other stakeholders who have a vested interest in your organization's sustainability performance. By keeping your employees informed, you can foster a sense of ownership and accountability for sustainability within your organization.

In short, communicating your sustainability efforts and progress transparently is an essential component of being a sustainable organization. By doing so, you can build trust, credibility, and support for your sustainability journey, which is critical for the long-term success of your sustainability initiatives.

Clearly Articulate Your Company's Sustainability Initiatives, Goals, & Achievements

As businesses increasingly recognize the importance of sustainability, it has become more critical than ever to articulate their sustainability initiatives, goals, and achievements clearly. A well-crafted sustainability message can help establish trust among customers, investors, and other stakeholders by demonstrating a company's commitment to environmental and social responsibility.

To ensure that your sustainability message resonates with your audience, it is essential to make it concise, consistent, and aligned with your overall brand identity. This means taking the time to understand your target audience and their values and tailoring your message to meet their expectations. Additionally, it is important to consider the tone and language used in your message to ensure that it is persuasive and compelling.

One of the key benefits of a well-articulated sustainability message is that it can help differentiate your brand from competitors in the marketplace. By highlighting your unique sustainability initiatives and achievements, you can establish yourself as a leader in your industry and attract customers who value sustainability.

When communicating your sustainability message, it is also important to be transparent about your progress towards your sustainability goals. This can help build trust with stakeholders and demonstrate your commitment to continual improvement. Whether you are reducing your carbon footprint, implementing sustainable sourcing practices, or investing in renewable energy, be sure to clearly communicate your progress and any challenges you may be facing.

Select The Appropriate Communication Channels To Reach Your Target Audience

To effectively communicate your company's sustainability initiatives and achievements, it's essential to consider the appropriate channels that will resonate best with your target audience. These channels include the company website, social media platforms, press releases, annual reports, sustainability reports, and industry events. Each channel offers a unique way to engage with stakeholders and convey your message, so tailoring your message to fit each format and audience is crucial.

- **Company Website**

Your company website serves as a centralized hub for information about your organization, including its sustainability efforts. Ensure that your website has a dedicated section for sustainability, where visitors can find detailed information about your sustainability strategy, goals, progress, and achievements. This section can include multimedia elements such as videos, infographics, and interactive content to engage users effectively. Consider creating a blog to provide regular updates and insights into your sustainability journey. Optimize your website for search engines (SEO) to improve its visibility and reach.

- **Social Media Platforms**

Social media platforms like LinkedIn, Twitter, Facebook, and Instagram are powerful tools for reaching a diverse audience and amplifying your message. Consider the demographics and preferences of your target audience when choosing which platforms to focus on. Share visually appealing content, such as images and videos, to increase engagement. Use hashtags and relevant keywords to expand your reach and connect with sustainability-focused communities. Encourage two-way communication by responding to comments, messages, and mentions promptly.

- **Press Releases**

Press releases are an effective way to communicate significant sustainability milestones, initiatives, and partnerships. Craft compelling and concise press releases that highlight the impact of your efforts and their relevance to current events or industry trends. Include quotes from key stakeholders, such as executives or partners, to add credibility and authenticity to your message.

Distribute your press releases through reputable wire services and media outlets to reach a broader audience.

- **Annual Reports**

Annual reports provide stakeholders, investors, and the public with a comprehensive overview of your company's financial and non-financial performance. Integrate sustainability reporting into your annual report to demonstrate your commitment to transparency and accountability. Include key sustainability metrics, targets, achievements, and initiatives, along with insights into how sustainability is integrated into your business strategy and operations. Use data visualization techniques to make complex information more accessible and engaging.

- **Sustainability Reports**

Standalone sustainability reports provide a deeper dive into your company's sustainability strategy, performance, and impacts. Tailor your sustainability report to your audience, whether it's investors, customers, or employees, by focusing on the most relevant topics and metrics. Use storytelling to bring your sustainability journey to life and highlight the human impact of your initiatives. Incorporate multimedia elements, such as case studies, interviews, and interactive graphics, to enhance engagement and understanding.

- **Industry Events**

Participating in industry events, conferences, and trade shows offers opportunities to showcase your sustainability initiatives, thought leadership, and best practices. Consider speaking engagements, panel discussions, and workshops to share your expertise and insights with a broader audience. Use event-specific communication channels, such as event websites, newsletters, and social media platforms, to promote your participation and engage with attendees before, during, and after the event.

Patagonia, an outdoor clothing and gear company, has a dedicated section on its website called "Environmental + Social Initiatives," which provides comprehensive information about its sustainability efforts. The company uses its social media platforms, particularly Instagram, to showcase its environmental activism and highlight its sustainability projects, such as its Worn Wear program[xc], which promotes garment repair and reuse. Patagonia also releases an annual environmental and social responsibility report that details its progress and challenges. The company participates in industry events and conferences, such as the Sustainable Apparel Coalition's Annual Members Meeting[xci], where it shares its sustainability practices and advocates for industry-wide change.

Tesla,[xcii] an electric vehicle and clean energy company, frequently uses press releases to announce significant sustainability milestones, such as achieving certain production or delivery targets or unveiling new sustainable products and initiatives. The company includes information about its sustainability efforts in its annual report, which is publicly available and outlines its financial and non-financial performance. Tesla publishes a standalone sustainability report, where it details its environmental and social impacts and discloses relevant metrics, such as greenhouse gas emissions and workplace diversity statistics. The company is also a frequent participant in industry events and trade shows, where it showcases its latest technologies and advocates for sustainable transportation and energy solutions.

Nestlé, a leading global food and beverage company, showcases its sustainability efforts through its "Nestlé in Society"[xciii] section on its website. Here, Nestlé outlines its commitments to sustainability, including goals like achieving net-zero greenhouse gas emissions by 2050 and using 100% renewable electricity by 2025. The company actively communicates its progress on social media platforms like Twitter and LinkedIn, sharing updates on initiatives such as water stewardship, sustainable sourcing, and packaging innovations.[xciv] Nestlé's annual sustainability report provides detailed insights into its environmental and social performance, demonstrating transparency and accountability to stakeholders. Through these channels, Nestlé engages with stakeholders, addresses inquiries, and collaborates on sustainable solutions.

These examples demonstrate how companies can effectively leverage various communication channels to reach their target audience with their sustainability initiatives. By using a combination of digital and traditional communication channels, companies can increase engagement, raise awareness, and drive positive change towards a more sustainable future.

Use Storytelling To Bring Your Sustainability Efforts to Life

Storytelling is a powerful tool in sustainability communication. It helps to humanize complex issues, engage stakeholders on an emotional level, and highlight the tangible impacts of sustainability initiatives.

Start by identifying key sustainability initiatives, projects, or success stories within your organization. These could be specific actions taken to reduce waste, improve energy efficiency, enhance community engagement, or empower employees. Choose stories that are authentic, impactful, and align with your company's values and goals.

Develop a narrative that clearly outlines the problem or challenge, the solution or initiative, and the resulting impact. Frame the story in a way that resonates with your target audience and emphasizes the positive outcomes of your sustainability efforts. Consider using a storytelling framework like The Hero's Journey to structure your narrative.

Visuals such as photos, videos, infographics, or animations can enhance your storytelling by making it more compelling and memorable. Use visuals that are authentic and relevant to the story you're telling. For example, if you're highlighting a waste reduction project, include images or videos of employees sorting and recycling materials.

Use storytelling to showcase the successes and impact of your sustainability efforts. Highlight specific metrics or data points that illustrate the positive outcomes of your initiatives, such as the amount of waste diverted from landfills, energy saved, or community members engaged.

Involve stakeholders in your storytelling by featuring their perspectives, experiences, or contributions to your sustainability initiatives. This can include employees, customers, partners, or community members who have been impacted by your sustainability efforts. Their stories can add authenticity and credibility to your narrative.

Ensure your storytelling is accessible to a wide range of audiences. Use language that is clear, concise, and jargon-free. Consider translating your stories into multiple languages if needed. Make sure your stories are easy to find and share, whether through your website, social media channels, or other communication platforms.

Storytelling is not a one-time effort; it should be an ongoing part of your sustainability communication strategy. Regularly update and share new stories that reflect the progress and evolution of your sustainability initiatives. This demonstrates your organization's commitment to sustainability and keeps stakeholders engaged and informed.

By incorporating storytelling into your sustainability communication strategy, you can effectively convey the importance of sustainability, engage stakeholders, and inspire action towards a more sustainable future.

However, it is important to note that incorporating storytelling into your sustainability communication strategy is not just about painting a better picture of your brand; rather, it's a powerful tool for raising awareness and fostering genuine engagement around sustainability issues. It's essential to use storytelling authentically, ensuring that narratives are grounded in truth and transparency to avoid the risk of greenwashing. By sharing authentic stories that highlight both the successes and challenges of your sustainability journey, you can build trust with stakeholders and demonstrate a sincere commitment to making meaningful change. Storytelling can humanize complex sustainability issues, making them relatable and inspiring action towards a more sustainable future. Through storytelling, organizations can connect with their audiences on an emotional level, driving awareness, empathy, and, ultimately, positive change.

Apple's "Mother Nature" advertisement[xcv] is a prime example of how storytelling can both captivate and divide audiences when it comes to sustainability messaging. The video effectively uses humor and a captivating storyline to convey a message about Apple's commitment to environmental responsibility. However, the advertisement received mixed reactions, with some viewers praising its storytelling approach while others criticized it for potential greenwashing.

On one hand, supporters of the ad appreciated its engaging narrative, which depicted Apple's efforts to prioritize sustainability and address environmental challenges. The playful interactions between the Apple employee and Mother Nature added a relatable and human element to the message, making it more memorable and impactful. Additionally, the ad showcased Apple's achievements in sustainability, such as transitioning to renewable energy and reducing carbon emissions, which resonated positively with environmentally-conscious consumers.[xcvi]

On the other hand, critics argued that the advertisement fell short of addressing the full scope of Apple's environmental impact and the challenges it faces in achieving sustainability goals.[xcvii] While the ad highlighted Apple's accomplishments, it glossed over potential areas of improvement or environmental controversies associated with the company's operations. Some viewers accused Apple of greenwashing, suggesting that the ad portrayed an overly positive image of the company's environmental efforts without acknowledging its shortcomings.

Overall, Apple's "Mother Nature" advertisement serves as a noteworthy example of the complexities involved in using storytelling for sustainability communication. While the ad successfully engaged audiences with its creative approach, it also sparked debate about the balance between highlighting achievements and addressing challenges transparently in sustainability messaging.

Foster Dialogue With Your Stakeholders & Encourage Two-Way Communication

Establishing an open and transparent dialogue with stakeholders is a cornerstone of successful sustainability communication. By fostering two-way communication, you can not only share your organization's sustainability efforts but also gain valuable insights, build trust, and foster a sense of shared responsibility.

Start by identifying your key stakeholders, including customers, employees, investors, suppliers, and local communities. Understand their interests, concerns, and expectations regarding your sustainability efforts. This will help you tailor your communication approach to each stakeholder group.

Provide multiple channels for stakeholders to engage with you. This could include dedicated email addresses, online forums, feedback forms on your website, or social media platforms. Ensure that these channels are easily accessible and user-friendly.

Actively seek feedback from stakeholders on your sustainability initiatives. This could involve conducting surveys, holding focus groups, or hosting community meetings. Encourage stakeholders to share their thoughts, suggestions, and concerns.

Be responsive to stakeholder inquiries related to your sustainability efforts. Respond promptly to questions and concerns, providing accurate and transparent information. If there are issues or challenges, acknowledge them and explain how you're working to address them.

Provide stakeholders with information about your sustainability goals, initiatives, and progress. This could include sharing reports, case studies, or success stories. Help stakeholders understand how their support and involvement contribute to your organization's sustainability journey.

Recognize and acknowledge the contributions of stakeholders to your sustainability efforts. Highlight examples of how stakeholders, such as employees, customers, or local communities, have made a positive impact. This fosters a sense of ownership and pride among stakeholders.

Encourage open dialogue with stakeholders by hosting regular meetings, webinars, or town halls. Use these opportunities to discuss your sustainability strategy, share updates, and gather input. Actively listen to stakeholder perspectives and incorporate their feedback into your decision-making process.

Use stakeholder feedback to improve your sustainability initiatives and communication approach continuously. Identify areas for improvement, address any concerns, and update stakeholders on the progress. This demonstrates your commitment to listening and responding to stakeholder input.

Be transparent about your sustainability performance and challenges. Share both successes and areas for improvement with stakeholders. This builds trust and credibility and shows that you are accountable for your actions.

By fostering dialogue with stakeholders, you can create a culture of transparency, trust, and collaboration around your sustainability efforts. This not only strengthens your relationships with stakeholders but also contributes to the overall success of your sustainability initiatives.

Regularly Measure & Report on Your Sustainability Performance

As sustainability moves to the forefront of corporate responsibility, it has become increasingly important for businesses to measure and report on their sustainability performance. This involves using relevant metrics and key performance indicators (KPIs) to provide transparent and credible data that demonstrates the impact of their sustainability initiatives and progress towards their sustainability goals.

Regularly measuring and reporting on sustainability performance not only allows businesses to track their progress but also helps them identify opportunities for improvement. By analyzing sustainability data, businesses can better understand their environmental impact as well as the social and economic implications of their operations. This information can then be used to inform decision-making and drive positive change across the organization.

In addition to tracking progress towards sustainability goals, measuring and reporting on sustainability performance can also help businesses build credibility with stakeholders. By providing transparent and reliable data, businesses can demonstrate their commitment to sustainability and build trust with customers, investors, employees, and other stakeholders.

Overall, measuring and reporting on sustainability performance is a critical component of corporate responsibility in today's business landscape. By using relevant metrics and KPIs to provide transparent and credible data, businesses can not only track their progress towards sustainability goals but also build credibility with stakeholders and drive positive change across the organization.

One notable example of a company regularly measuring and reporting on its sustainability performance is Unilever. Unilever has been at the forefront of sustainability efforts within the corporate sector. The company has established ambitious sustainability goals[xcviii], such as achieving net-zero emissions from all its products by 2039 and ensuring that all its packaging is reusable, recyclable, or compostable by 2025.

To track its progress towards these goals and other sustainability targets, Unilever regularly measures and reports on its sustainability performance using a comprehensive set of metrics and KPIs. The company's Sustainable Living Plan, launched in 2010, outlines specific targets across various sustainability pillars, including environmental impact, social responsibility, and economic growth.

Unilever's annual Sustainable Living Report[xcix] provides transparent and credible data on its sustainability performance, highlighting achievements, challenges, and areas for improvement. The report covers key areas such as greenhouse gas emissions, water usage, waste reduction, sustainable sourcing, and social impact initiatives.

By consistently measuring and reporting on its sustainability performance, Unilever not only tracks its progress but also holds itself accountable to stakeholders. The company's commitment to transparency and accountability has helped build trust with consumers, investors, and other stakeholders, reinforcing its reputation as a leader in corporate sustainability.

Avoid Greenwashing & Greenhushing

As consumers become more environmentally conscious, companies have started adopting eco-friendly practices to appeal to this growing market. However, not all companies are genuine in their approach and engage in practices known as greenwashing.

Greenwashing is the act of making false or exaggerated claims about the environmental benefits of a product or service. This can mislead consumers into thinking that they are making a sustainable choice when, in reality, they are not.

To avoid greenwashing, companies should strive to be transparent about their environmental practices. They should provide clear and accurate information about the materials used, the manufacturing process, and the environmental impact of their products. This will help consumers make informed decisions and hold companies accountable for their actions.

Also, regulators are enhancing their focus on the topic. Only recently has the European Parliament given its final green light to a directive that will improve product labelling and ban the use of misleading environmental claims. The directive adopted seeks to protect consumers from misleading marketing practices and help them make better purchasing choices. To achieve this, a number of problematic marketing habits related to greenwashing and the early obsolescence of goods will be added to the EU list of banned commercial practices.[c]

Most importantly, the new rules aim to make product labelling clearer and more trustworthy by banning the use of general environmental claims like "environmentally friendly", "natural", "biodegradable", "climate neutral" or "eco" without proof.

The use of sustainability labels will also now be regulated, given the confusion caused by their proliferation and failure to use comparative data. In the future, only sustainability labels based on official certification schemes or established by public authorities will be allowed in the EU.

Additionally, the directive will ban claims that a product has a neutral, reduced, or positive impact on the environment because of emissions-offsetting schemes.

Hence, strong governance practices play a pivotal role in preventing greenwashing. Having multiple stakeholders involved, especially at the senior management level, ensures that environmental claims are not only accurate but also aligned with the company's actual sustainability efforts. This approach fosters checks and balances within the organization, reducing the likelihood of misleading or exaggerated environmental claims.

Engaging stakeholders, such as environmental experts, regulatory bodies, consumer advocacy groups, and even independent auditors, can further enhance the credibility of a company's environmental practices. These stakeholders provide valuable insights, conduct assessments, and verify the accuracy of environmental claims, thereby strengthening the company's commitment to transparency and authenticity in sustainability efforts.

In essence, combining transparent communication with strong governance practices involving multiple stakeholders at senior management levels is key to avoiding greenwashing and building trust with consumers, investors, and the wider community. This approach not only upholds ethical standards but also contributes to meaningful environmental impact mitigation and sustainability progress.

The rebranding of the **World Coal Association** to "FutureCoal: The Global Alliance for Sustainable Coal" has raised eyebrows and sparked skepticism within the sustainability community.[ci] Critics question the credibility of labelling coal, a significant contributor to man-made climate change, as "sustainable." The move has been met with incredulity, with some likening it to an April Fool's joke arriving months ahead of schedule.

The assertion by FutureCoal that the coal industry can be sustainable is met with skepticism by experts and observers alike. Bitasta Roy Mehta, a human resources practitioner, finds the notion of labelling coal as green almost comical, drawing a parallel with the absurdity of branding Coca-Cola as a health drink. This sentiment is echoed by sustainability consultant Klemen Risto Bizjak, who remarks that the concept of "sustainable coal" defies logic. He suggests that the term may be an attempt to frame discussions around the sustainability of coal production, particularly amidst lobbying efforts by fossil fuel associations at global climate conferences like COP28.

The rebranding of the World Coal Association as FutureCoal highlights a prevalent issue known as greenwashing, where companies or industries attempt to portray themselves as environmentally friendly or sustainable despite evidence to the contrary. By associating coal, a fossil fuel known for its environmental impact and contribution to climate change, with sustainability,

FutureCoal risks undermining genuine efforts to transition towards cleaner energy sources and mitigate the effects of climate change.

The European Consumer Organization, along with environmental nonprofits Client Earth and ECOS, have accused **Coca-Cola**, **Danone**, and **Nestlé** of making misleading claims regarding their single-use plastic water bottles.[ci] These claims assert that the bottles are either "100 percent recycled" or "100 percent recyclable." However, the complaint lodged with the European Commission argues that the packaging from these companies is never entirely made from recycled materials, and their recyclability depends on various factors, such as the availability of waste management infrastructure, particularly in Asian countries where such infrastructure may be lacking.

Rosa Pritchard, a plastics lawyer at **ClientEarth**, emphasized that achieving a "100 percent" recycling rate for bottles is technically unfeasible. Additionally, she pointed out that even if bottles are made with recycled plastic, they can still have harmful impacts on both people and the planet. This criticism highlights a common challenge in the packaging industry, where claims of sustainability may not always align with the actual environmental impact of the products.

In response to the allegations, Nestlé stated its commitment to reducing the use of plastic packaging, while Danone affirmed its ongoing investment in recycling infrastructure. Coca-Cola defended its claims by asserting that they can be substantiated. However, the controversy surrounding these claims underscores the importance of transparency and accuracy in sustainability marketing. It also serves as a reminder of the complexities involved in achieving genuine sustainability in the consumer goods sector, particularly regarding packaging materials and their environmental implications.

Another practice to avoid is greenhushing. This refers to the act of withholding information about environmental practices, often due to fear of negative publicity or legal action. Companies should not hide behind a veil of secrecy when it comes to their environmental practices. Instead, they should be open and willing to engage in dialogue with consumers and stakeholders.

In short, companies should be transparent and honest about their environmental practices to avoid greenwashing and greenhushing. By doing so, they can build trust with consumers, who are increasingly prioritizing sustainability in their purchasing decisions.

Conclusion

In conclusion, communication and transparency are pivotal in shaping the narrative of your sustainability journey and fostering support from your stakeholders. As the world grapples with complex environmental and social challenges, companies are increasingly expected to not only have sustainability initiatives but also communicate them effectively.

Start by clearly articulating your sustainability initiatives, goals, and achievements. This not only ensures alignment with your company's brand identity but also helps in making your message concise and consistent. Additionally, leveraging appropriate communication channels such as company websites, social media platforms, press releases, annual reports, sustainability reports, and industry events is crucial. Ensure that the message is tailored to fit the format and audience of each channel.

Using storytelling as a tool to bring your sustainability efforts to life can significantly enhance engagement and understanding. Highlighting specific projects, success stories, and the tangible impact of your initiatives on the environment, community, and stakeholders can make the narrative relatable and compelling. Integrate visuals such as photos and videos to make your stories more engaging and impactful.

Encouraging dialogue with stakeholders is a cornerstone of building trust and support. This involves fostering two-way communication by inviting feedback, answering questions, and addressing concerns related to your sustainability efforts. Regularly measuring and reporting on your sustainability performance using relevant metrics and key performance indicators (KPIs) is vital. Provide transparent and credible data to demonstrate the impact of your initiatives and progress towards your sustainability goals.

Finally, it is essential to avoid greenwashing and greenhushing, both of which undermine the credibility of sustainability efforts. Greenwashing involves misleadingly portraying a company as environmentally friendly when it is not, while greenhushing involves hiding negative environmental performance. By being transparent and accountable, you can build trust and credibility and ultimately drive meaningful change towards a more sustainable future.

Chapter 10
Collaboration & Partnerships

"Collaboration and partnerships are the cornerstone of driving meaningful change towards sustainability. By working together, organizations can leverage collective expertise and resources to address complex challenges and create lasting positive impact for people and the planet."

In the pursuit of a more sustainable and responsible business model, collaboration and partnerships play a pivotal role in amplifying the impact of sustainability initiatives and contributing to broader systemic change. By fostering strategic collaborations with other organizations, industry groups, and non-governmental organizations (NGOs), your company can harness collective expertise, resources, and influence to drive meaningful progress towards sustainability goals. This involves seeking out opportunities for joint initiatives, shared research and development, and cross-industry partnerships to address common challenges and maximize the effectiveness of sustainability efforts. Additionally, by actively participating in industry-wide initiatives and platforms, your company can help shape collective action, share best practices, and advance industry standards for sustainability, ultimately contributing to a more sustainable and responsible business ecosystem.

Sustainability initiatives often have significant dependencies that extend beyond the boundaries of any single organization. To achieve meaningful impact, collaboration is not just desirable but essential. By partnering with a diverse array of stakeholders, organizations can tap into a wealth of collective expertise and resources that can vastly enhance the effectiveness of their sustainability efforts.

Collaboration allows organizations to access the collective knowledge, perspectives, and insights of partners, stakeholders, and other entities working in the sustainability space. This enables them to develop more comprehensive and robust ESG (Environmental, Social, and Governance) strategies that are better aligned with industry best practices and stakeholder expectations.

Through collaboration, companies can address complex sustainability challenges that may be beyond their individual capacities. They can leverage each other's strengths, networks, and resources to innovate and implement solutions that have a more significant and lasting impact on the environment, society, and economy. Ultimately, collaboration fosters a collective approach to sustainability, which is vital for driving systemic change and achieving shared goals.

Addressing Complexity Through Collaboration

Sustainability, particularly in the context of environmental conservation, social responsibility, and ethical governance, encompasses multifaceted and interdependent issues. These include climate change, social inequality, and supply chain sustainability, all of which present complex challenges that transcend individual organizational boundaries. Embracing collaboration allows companies to tackle these intricacies head-on.

Pooling Resources: Collaboration enables companies to combine their resources, whether they are financial, technical, or intellectual. By sharing the burden, organizations can start more ambitious sustainability initiatives, including research and development, community projects, or large-scale investments in green technology.

The **Sustainable Apparel Coalition** (SAC) is a prime example of how companies in the apparel and textile industry are pooling their resources to address complex sustainability challenges. Members of the SAC, including major brands like Nike, Adidas, and H&M, share data, tools, and best practices to improve environmental and social performance throughout the supply chain. This collaboration has led to the development of tools like the Higg Index, a comprehensive sustainability assessment tool used by thousands of companies to measure and benchmark their sustainability performance.[cii]

Sharing Best Practices: By working together, companies can exchange knowledge and insights on sustainable practices that have proven successful in their respective industries. This cross-pollination of ideas and experiences fosters a culture of continuous improvement and innovation, enabling organizations to adopt more effective strategies for sustainability.

The Renewable Energy Buyers Alliance (REBA) is a collaboration of companies like Facebook, Google, and Walmart that are working together to increase the use of renewable energy in the corporate sector. Through initiatives like the Corporate Renewable Energy Buyers' Principles, REBA members share best practices, policies, and strategies for purchasing renewable energy. This collaboration has helped drive the growth of renewable energy procurement, making it more accessible and cost-effective for companies of all sizes.[ciii]

Engaging In Collective Action: Collaboration empowers companies to engage in collective action, where multiple entities come together to address common sustainability goals or challenges. This could involve initiatives like industry-wide emissions reduction programs, joint community development projects, or advocacy for regulatory changes that support sustainable business practices.

The **Climate Leadership Council** (CLC)[civ] is a coalition of major corporations, environmental organizations, and opinion leaders that have come together to advocate for a carbon tax as a solution to climate change.[cv] Companies like BP, ExxonMobil, and Microsoft are part of this coalition, which advocates for a carbon tax that would incentivize companies to reduce their emissions while returning the revenue to American households. This collaborative effort has brought together diverse stakeholders to push for a policy that could significantly reduce greenhouse gas emissions.

Addressing interconnected challenges such as climate change and social inequality requires a coordinated and collective effort. Through collaboration, companies can leverage their collective influence and resources to drive meaningful change and create a more sustainable and equitable world.

One notable example of collaborative efforts is the **Net Zero Asset Owners Alliance**, an initiative launched by the UN-convened Net Zero Asset Owner Alliance.[cvi] This alliance brings together institutional investors who are committed to transitioning their investment portfolios to net-zero greenhouse gas emissions by 2050. By collaborating under this initiative, asset owners align their investment strategies with climate goals, engage with portfolio companies to reduce emissions and promote sustainable business practices across industries. This collective action demonstrates how companies can work together to address climate change effectively.

Similarly, various public-private partnerships facilitated by organizations like the UN aim to tackle social inequality and promote inclusive growth. These partnerships often involve governments, businesses, nonprofits, and community organizations working together on initiatives related to education, healthcare, job creation, and social welfare. By pooling resources, expertise, and networks, these collaborations can have a broader and more sustainable impact on addressing systemic issues that contribute to social inequality.

Strengthening Stakeholder Engagement Through Collaboration

Building and nurturing relationships with stakeholders is essential for sustainable growth and success. By partnering and collaborating with a wide range of stakeholders, companies can demonstrate their commitment to effective stakeholder engagement, showing that they value and prioritize the needs and perspectives of investors, customers, employees, and communities.

Investor Confidence: Collaboration and partnerships in sustainability initiatives can significantly impact investor confidence. When companies actively engage with stakeholders and work together to address environmental, social, and governance (ESG) challenges, it signals a commitment to long-term value creation and risk mitigation. This can be particularly compelling for impact-driven investors and those seeking sustainable and responsible investment opportunities.

Customer Loyalty: A company that collaborates on sustainability initiatives can differentiate itself from competitors and enhance its brand image in the eyes of consumers. In an increasingly socially conscious market, consumers are more likely to support companies that actively engage in meaningful sustainability partnerships. This can lead to increased customer loyalty, brand advocacy, and, ultimately, higher sales.

Employee Morale & Retention: Collaborative sustainability efforts can also have a positive impact on employee morale and retention. When employees see their company working with others to address societal and environmental challenges, it fosters a sense of pride and purpose. This can contribute to higher employee engagement, retention rates, and a more motivated workforce.

Community Relations: Lastly, collaboration and partnerships can strengthen relationships with local communities where businesses operate. By engaging with community organizations, non-profits, and other local stakeholders, companies can gain valuable insights, build trust, and enhance their social license to operate. This can help prevent or mitigate potential conflicts and ensure that business activities align with community needs and expectations.

Partnerships and collaborations in sustainability not only demonstrate a company's commitment to stakeholder engagement but can also lead to tangible benefits such as enhanced investor confidence, customer loyalty, employee morale, and positive community relations. By working together, businesses can create more meaningful and impactful solutions to global challenges, ultimately driving sustainable growth and prosperity for all stakeholders.

Fostering Innovation Through Cross-Sector Collaboration

When companies collaborate with partners from different sectors and industries, they can unleash a wave of innovation and creativity. These partnerships bring together diverse perspectives, expertise, and resources, providing fertile ground for the development of novel solutions to sustainability challenges.

Through collaboration, businesses can co-create innovative products and services that address pressing environmental and social needs. For example, a partnership between a technology company and a sustainability-focused NGO might result in the creation of a new app that helps consumers reduce their carbon footprint. Similarly, an alliance between a pharmaceutical company and a conservation organization might lead to the development of sustainable healthcare products derived from natural resources.

Cross-sector collaboration can also spur the development of new business models that prioritize sustainability. For instance, companies in the fashion industry may collaborate with recycling firms to create circular supply chains where products are designed to be reused, recycled, or repurposed. This can reduce waste and promote a more sustainable approach to production and consumption.

Collaborating with partners from different sectors can also help accelerate the adoption of technological solutions to ESG challenges. For example, a collaboration between an automotive manufacturer and a renewable energy company might lead to the widespread adoption of electric vehicles powered by clean energy. By leveraging each other's expertise and capabilities, companies can overcome technological barriers and drive innovation at scale.

One notable example of cross-sector collaboration driving innovation is the partnership between Tesla, a leading electric vehicle manufacturer, and Panasonic, a global electronics company. Together, they developed and produced lithium-ion batteries[cvii] for Tesla's electric vehicles, revolutionizing the automotive industry and paving the way for the widespread adoption of electric mobility.

In short, cross-sector collaboration is a powerful catalyst for innovation, enabling companies to develop new sustainable products, services, and business models, as well as accelerate the adoption of technological solutions to ESG challenges. By partnering with organizations from different sectors and industries, businesses can tap into a wealth of expertise and resources, driving positive change and advancing sustainability on a global scale.

Amplifying Impact Through Collaborative Partnerships

Collaborating with other organizations offers companies the opportunity to amplify their impact on environmental, social, and governance (ESG) issues. These partnerships enable the scaling of successful initiatives, the replication of best practices, and the mobilization of resources for large-scale projects.

When companies collaborate with partners, they can expand the reach and impact of their successful sustainability initiatives. For instance, a local community project aimed at reducing plastic waste could be scaled up to a national level through collaboration with government agencies and NGOs. This allows the initiative to have a broader impact and reach a larger audience.

Partnerships also enable the replication of best practices across different industries and regions. For example, a multinational corporation may share its successful water conservation practices with suppliers in developing countries, leading to the widespread adoption of sustainable water management practices.

Collaborative partnerships can bring together the financial, technical, and human resources needed to undertake large-scale sustainability projects. For instance, a coalition of companies, NGOs, and government agencies may collaborate on a reforestation initiative, pooling their resources to plant millions of trees and combat deforestation.

The Harvard Business Review, for example, explains how the **Sustainable Food Lab**[cviii]—an organization created through a collaboration between businesses, NGOs, and government agencies—has been successful in promoting sustainable agricultural practices. Similarly, the **Zero Waste International Alliance** (ZWIA)[cix,cx] is a global network that encourages waste reduction and promotes the principles of zero waste through collaboration between organizations from different sectors.

In essence, collaborative partnerships offer companies the opportunity to amplify their impact on ESG issues by scaling successful initiatives, replicating best practices, and mobilizing resources for large-scale projects. By working together with other organizations, businesses can achieve greater collective impact and drive positive change on a global scale.

Conclusion

Collaboration and partnerships are essential pillars for achieving meaningful and lasting sustainability impacts. However, they are only fruitful when they are guided by clear objectives, transparent communication, and a shared commitment to creating positive change.

Setting specific goals for collaboration is crucial. By doing so, companies can ensure that their efforts align with their overall sustainability strategy and that all parties involved have a clear understanding of the desired outcomes. Moreover, companies should strive to create win-win-win situations, benefiting themselves, their partners, and the broader sustainability cause. This approach not only enhances the effectiveness of the partnership but also ensures that it remains sustainable over time.

Transparency and open communication are the bedrock of successful partnerships. Companies must be willing to share relevant information, ideas, and resources with their partners. This level of openness fosters trust and fosters stronger relationships between collaborators. Moreover, it allows for the development of innovative solutions to complex sustainability challenges that would be difficult to achieve individually.

However, the work doesn't stop once a partnership is established. Companies must continually evaluate their collaborations to ensure they are creating the intended impact. Regular assessments help identify areas for improvement and allow for adjustments to be made, ensuring that partnerships remain effective in driving meaningful and lasting change.

An exemplary case of effective collaboration and sustainable partnership is the "Fashion for Good"[cxi] initiative, a global platform that aims to make fashion more sustainable. Founded by the C&A Foundation, it brings together brands, manufacturers, innovators, and investors to accelerate the adoption of sustainable practices in the fashion industry. Through various programs, such as the Good Fashion Fund[cxii] and the Accelerator Program, Fashion for Good is creating a collaborative ecosystem that fosters innovation, drives change, and sets new standards for sustainability in the fashion sector. This initiative exemplifies how strategic partnerships, guided by clear objectives and a shared vision, can lead to significant and transformative impacts on sustainability.

Why, What and How?

"Sustainability is not a destination but a continuous journey. The journey towards sustainability is not just about what we achieve today but how we shape tomorrow. It's a continuous commitment to balance prosperity with responsibility, ensuring a thriving future for generations to come."

ESG and sustainability have emerged as some of the most significant challenges and opportunities for organizations worldwide. In today's increasingly interconnected world, companies are not only expected to deliver financial performance but also demonstrate their commitment to ESG factors. These include climate change, human rights, diversity and inclusion, and ethical business practices, among others. As organizations grapple with these complex issues, there is a growing recognition that sustainability is no longer an optional endeavor but a fundamental aspect of business strategy. It requires companies to integrate sustainability principles into their core operations, supply chains, and decision-making processes.

Sustainability has a profound impact on companies' reputation and bottom-line performance. A company's commitment to ESG factors such as environmental stewardship, social responsibility, and strong governance practices can significantly influence how it is perceived by stakeholders, including customers, investors, employees, and the broader community.

Firstly, embracing sustainability can enhance a company's reputation and brand image. Consumers today are increasingly conscious of environmental and social issues, and they prefer to support businesses that align with their values. Companies that demonstrate genuine efforts towards sustainability, such as reducing carbon emissions, implementing fair labor practices, or supporting community initiatives, often earn trust and loyalty from customers. This positive reputation can translate into increased customer engagement, brand loyalty, and, ultimately, higher sales and market share.

Secondly, sustainability initiatives can directly impact a company's bottom line in various ways. For instance, adopting energy-efficient practices or transitioning to renewable energy sources can lead to cost savings and operational efficiencies over time. Also, implementing sustainable supply chain practices, such as sourcing ethically produced materials or reducing waste, can lower procurement costs and improve resource utilization. Similarly, focusing on sustainability can increase revenues, as significant investments are required in sustainability, providing opportunities for companies to support this transition through technologies, products, and services.

Furthermore, investors and financial institutions are increasingly considering ESG performance as a key factor in their investment decisions. Companies with strong ESG credentials are seen as more resilient, forward-thinking, and capable of managing risks effectively, which can attract investment capital and lower borrowing costs. Conversely, companies that neglect ESG considerations may face reputational damage, regulatory scrutiny, and financial risks, such as fines, legal liabilities, or divestment by ESG-focused investors.

Defining the sustainability journey is a crucial step for organizations. It involves setting clear objectives and goals, identifying material issues, and establishing a roadmap for implementation. This journey is unique to each organization, depending on its size, industry, location, and stakeholder expectations. It requires organizations to align their sustainability efforts with their business strategy, corporate values, and long-term vision. It also necessitates ongoing dialogue and engagement with key stakeholders, including customers, employees, investors, and local communities.

For sustainability initiatives to succeed, every employee must understand their role in the organization's sustainability journey. This involves creating a culture of sustainability where employees are aware of the organization's sustainability goals and their individual responsibilities in achieving them. It requires providing adequate training, resources, and support to employees to enable them to contribute effectively to sustainability efforts. Employees should be empowered to identify opportunities for improvement, participate in sustainability initiatives, and integrate sustainability into their day-to-day work.

Employees play a crucial role not only in understanding and implementing sustainability initiatives but also in advocating for a greater focus on sustainability within their organizations. They can actively engage in discussions, provide feedback, and propose ideas to enhance sustainability practices. By voicing their concerns and demonstrating the importance of sustainability, employees can influence decision-makers and contribute to shaping a more sustainable organizational culture. Their advocacy can lead to the adoption of more ambitious sustainability goals, the allocation of resources towards sustainable practices, and the integration of sustainability considerations into strategic decision-making processes. Employees, as internal champions for sustainability, can drive meaningful change and inspire collective action towards a more sustainable future.

Overall, ESG and sustainability present both challenges and opportunities for organizations. While they require effort and resources, they also offer numerous benefits, including enhanced reputation, improved risk management, and increased innovation and competitiveness. Organizations that embrace sustainability as a strategic imperative and embed it into their corporate DNA are better positioned to thrive in the evolving business landscape. As the global community continues to prioritize sustainability, organizations must commit to meaningful action and ongoing improvement in their sustainability journey.

The sustainability journey can be likened to the digital journey that businesses have experienced over the last two decades. Twenty years ago, having a digital strategy was a differentiating factor for companies. Only a few organizations had embraced digital transformation, while others were still figuring out how to leverage digital technologies to improve their operations and customer experiences. Fast forward to today, and having a digital strategy is no longer an option—it's a necessity. Every company, regardless of its industry or size, has a digital strategy or plan in place. Digital technologies have become an integral part of business operations, from customer engagement and supply chain management to data analytics and decision-making.

Similarly, the sustainability journey is evolving in a similar manner. Just as companies recognized the importance of digital transformation and adapted to it, they are now realizing the significance of sustainability and ESG factors. What was once considered a niche area has now become a critical business imperative. Sustainability is no longer an optional endeavor but a fundamental aspect of business strategy. Companies are increasingly expected to demonstrate their commitment to environmental stewardship, social responsibility, and good governance.

In the future, having a sustainability journey or plan will be just as essential as having a digital strategy. It will not only be a means of meeting regulatory requirements or gaining a competitive advantage but also a way to secure the long-term viability of the business. In addition to being essential for meeting regulatory requirements and gaining a competitive advantage, having a sustainability journey or plan will be required by all stakeholders in the future. Employees will prefer employers who prioritize sustainability, as it aligns with their values and expectations for responsible corporate citizenship. Customers are increasingly willing to pay higher prices for sustainable products and support brands with strong environmental and social commitments. Regulators are enhancing their focus on ESG factors, making it imperative for businesses to comply with sustainability standards and reporting requirements. Investors are also placing greater importance on ESG considerations, integrating them into their investment decision-making processes to assess the long-term risks and opportunities associated with sustainability performance. Therefore, a robust sustainability strategy will not only benefit the business but also fulfil the expectations of employees, customers, regulators, and investors, ensuring the long-term viability and success of the organization.

Sustainability will shape how companies operate, innovate, and create value for all stakeholders. Those organizations that fail to embrace sustainability will face the risk of being left behind as consumers, investors, and regulators increasingly prioritize sustainable business practices.

So, it is safe to say that just as the digital journey has transformed businesses over the past two decades, the sustainability journey is poised to do the same in the coming years. Companies that proactively embrace sustainability and embed it into their core business strategies will be better positioned to thrive in the future business landscape. As the saying goes, "Sustainability is not a cost; it's an investment in the future."

When developing and implementing a sustainability strategy, there are several "do's" and "don'ts" to consider:

Do's:

- **Set Clear Objectives and Targets:** Establish clear, measurable, and achievable sustainability objectives and targets aligned with your business goals.

- **Integrate ESG into Business Strategy:** Embed sustainability considerations into your organization's core business strategy, making it an integral part of decision-making processes.
- **Collaborate Across Functions:** Involve representatives from various departments, including operations, finance, and marketing, to ensure a comprehensive and cross-functional approach to sustainability.
- **Engage with Stakeholders:** Regularly communicate and engage with stakeholders, including customers, employees, investors, and communities, to understand their expectations and concerns and incorporate their feedback into your strategy.
- **Focus on Materiality:** Identify and prioritize sustainability issues that are most material to your business and stakeholders to ensure that resources are allocated effectively.
- **Invest in Employee Training:** Provide training and awareness programs to employees to enhance their understanding of sustainability and encourage their involvement in sustainability initiatives.
- **Regular Monitoring and Reporting:** Establish a robust monitoring and reporting system to track progress against targets and communicate transparently with stakeholders.
- **Collaborate with Partners:** Seek collaboration with suppliers, industry peers, and other stakeholders to drive collective action and scale impact.

Don'ts:

- **Greenwashing:** Avoid overstating or exaggerating your sustainability achievements or making misleading claims about environmental or social impact.
- **Lack of Transparency:** Avoid withholding or providing incomplete information about your sustainability practices and performance.
- **Failure to Align with Business Strategy:** Don't treat sustainability as a separate initiative or an afterthought; it should be integrated into your overall business strategy.
- **Ignoring Stakeholders:** Don't overlook the expectations and concerns of stakeholders; their engagement is crucial for the success of your sustainability strategy.

- **Ignoring Risks and Opportunities:** Don't overlook the risks and opportunities associated with sustainability issues that may impact your business operations and performance.
- **Neglecting Measurement and Reporting:** Don't neglect the importance of accurate measurement, reporting, and validation of sustainability performance to ensure credibility and transparency.
- **Ignoring Industry Standards and Frameworks:** Don't ignore established sustainability standards and frameworks; adherence to these guidelines enhances the credibility and comparability of your reporting.

By adhering to these do's and don'ts, organizations can develop and implement a robust and effective sustainability strategy that drives positive environmental, social, and governance (ESG) impacts while contributing to long-term business success.

Embracing sustainability is about recognizing the triple bottom line principle that sustainability is beneficial for People, Planet, and Profit (PPP). This means that sustainability initiatives should aim to improve the well-being of employees, customers, and communities (People), reduce environmental footprints promote conservation efforts (Planet), and also generate economic value and enhance financial performance (Profit). By focusing on PPP, organizations can create value that extends beyond financial gains, fostering a holistic approach to business that integrates sustainability into every aspect of operations and decision-making.

ESG Glossary

Biodiversity: The variety of plant and animal life in the world or in a particular habitat/region. Within ESG, the term refers to biodiversity loss or ecosystem degradation.

Carbon footprint: The total amount of greenhouse gasses, primarily carbon dioxide (CO2), emitted directly or indirectly by an individual, organization, event, or product over its lifecycle. Measuring and reducing carbon footprints is essential for mitigating climate change.

Carbon offset: A financial instrument that represents a reduction in greenhouse gas emissions and can be traded on carbon markets. Offsets are used to compensate for emissions that cannot be reduced directly and are generated by projects such as renewable energy, energy efficiency, and reforestation.

Carbon pricing: Carbon pricing refers to the use of economic incentives, such as taxes or cap-and-trade systems, to encourage the reduction of greenhouse gas emissions. By putting a price on carbon emissions, carbon pricing provides an economic incentive for companies and individuals to reduce emissions and invest in low-carbon technologies.

Circular economy: An economic model that aims to minimize waste and maximize resource efficiency. It promotes the continuous use of resources through practices like recycling, reusing, and remanufacturing, rather than the traditional linear "take-make-dispose" model.

Climate change: Change in climate patterns and extreme weather conditions attributed to the emission of carbon dioxide into the atmosphere

Community Engagement: The active involvement of companies and organizations in the communities in which they operate, including through philanthropy, volunteering, and other forms of civic engagement.

COP – or Conference of the Parties: The COP is the supreme decision-making body of the UN Convention on Climate Change. All States that are Parties to the Convention are represented at the COP, at which they review the implementation of the Convention and any other legal instruments that the COP adopts and take decisions necessary to promote the effective implementation of the Convention, including institutional and administrative arrangements.

CSR- Corporate social responsibility: A type of international private business self-regulation that aims to contribute to societal goals of a philanthropic, activist, or charitable nature by engaging in or supporting volunteering or ethically-oriented practices.

Decarbonization: Decarbonization refers to the process of reducing the carbon intensity of energy systems and the economy as a whole. This can be achieved by low-carbon energy sources, energy efficiency, and other mitigation measure.

Diversity and inclusion: Diversity refers to the differences people may have in terms of their gender, age, race, religion, disability, or other characteristics. Inclusion is about celebrating diversity, challenging inequality, and ensuring people feel respected.

Environmental factors: The E in ESG includes assessing factors that have an impact on the environment, such as addressing climate change impact and solutions, energy efficiency, waste management, pollution, and the broader impacts that a company and its products have upon the planet.

Governance: The G in ESG is also known as corporate governance. Refers to the rights and responsibilities of stakeholders and how well a company is run. Includes issues such as executive pay, corruption and board diversity.

Green bonds: Bonds issued to fund a specific project with an environmental purpose.

Greenwashing: The promotion of a product through misleading claims as to its environmental credentials.

GRI – <u>Global Reporting Initiative</u>: an international organization that develops and maintains a sustainability reporting framework. Companies use this framework to disclose information about their environmental, social, and governance (ESG) performance. The GRI framework helps companies to be transparent about their sustainability practices and provides stakeholders with relevant and comparable information.

Human rights: Basic rights that belong to everyone, including the right to life, liberty, freedom from slavery and torture, and the freedom of opinion and expression. The UN Declaration on Human Rights is widely recognized as the benchmark for these basic standards.

Internal Controls: The processes and practices a company uses to ensure that its operations are efficient, effective, and compliant with applicable laws and regulations. Strong internal controls are important for ESG as they help ensure that a company is operating in a responsible and effective manner and that its activities align with stakeholders' interests.

IPCC – International Panel on Climate Change: A body created by the United Nations with the intention of providing scientific assessments on climate change, its impact, and future risks, as well as suggestions for mitigating impact and disruptions.

Materiality: Materiality is a measure of the importance of specific topics and information during the investment process. The more significant a topic is, the more material it is, and vice versa.

Net Zero: Achieving a balance between the greenhouse gases emitted into the atmosphere and those removed or offset, resulting in no net addition to the overall amount of greenhouse gases. This goal is crucial in combating climate change.

Renewable energy: Energy that is naturally replenished from sources such as solar, wind, water and geothermal heat.

Resilience: A measure of how well a building is prepared for potentially disruptive events and changing conditions, such as earthquake-proofing or features designed to combat negative effects from long-term risks like climate change.

Risk Management: the processes and practices a company uses to identify, assess, and manage potential risks to its business, such as environmental, social, or financial risks. Effective risk management is important for ESG as it helps ensure that a company is prepared to address potential risks and minimize their impact on its operations and stakeholders.

SASB – <u>Sustainability Accounting Standards Board</u>: a non-profit organization that provides sector-specific standards for sustainability reporting. **Science-based target:** Provides a roadmap for companies to future-proof growth by creating a roadmap of how much to reduce carbon emissions and how quickly the reduction needs to happen.

Scope 1 emissions: Scope 1 emissions refer to direct greenhouse gas (GHG) emissions that occur from sources owned or controlled by an organization. These emissions include those generated from on-site combustion of fossil fuels, such as emissions from company-owned vehicles or equipment, as well as emissions from industrial processes.

Scope 2 emissions: Scope 2 emissions refer to indirect greenhouse gas emissions that result from the consumption of purchased electricity, heat, or steam by an organization. These emissions occur outside the organization's boundaries but are associated with its activities. Scope 2 emissions are typically generated by external sources, such as power plants that supply electricity to the organization.

Scope 3 emissions: Greenhouse gas emissions that occur indirectly from the activities of an organization but are a consequence of its operations. Scope 3 emissions include all emissions from sources not owned or directly controlled by the organization, such as those from purchased goods and services, transportation, and waste disposal.

Shareholder activism: Shareholders looking to change a company's behavior. Increasingly used to describe those pushing for improvements to a company's ESG policy and practices.

Social factors: The S in ESG refers to areas such as diversity and inclusion, human rights, inequality, data protection and health and safety

Stakeholder Engagement: Refers to a company's communication and interaction with its stakeholders, including shareholders, employees, customers, and local communities, to understand and address their concerns and expectations. Stakeholder engagement is important for ESG as it helps companies identify potential ESG risks and opportunities and ensure that their activities align with the values and interests of their stakeholders.

TCFD – Task Force on Climate-related Financial Disclosures: a voluntary reporting framework that provides guidelines for disclosing financial risks and opportunities related to climate change.

TNFD- Taskforce on Nature-related Financial Disclosures: A set of disclosure recommendations and guidance that encourage and enable business and finance to assess, report and act on their nature-related dependencies, impacts, risks and opportunities.

Transparency: a company's openness and accessibility in sharing information about its operations, performance, and governance. Transparency is important for ESG as it helps ensure that a company is accountable and responsible in its activities and that stakeholders have access to information to make informed decisions.

Triple bottom line: A framework that measures organizational success based on three dimensions: economic, social, and environmental. It emphasizes the need for businesses to consider not only their financial outcomes but also outcomes related to society and the environment.

UN Sustainable Development Goals (UN SDGs): The United Nations set 17 core goals to encourage action to improve global issues such as social well-being, inequality and climate impact. The 2030 Agenda for Sustainable Development was adopted by UN member states in 2015.

Waste Management: The collection, transport, treatment, and disposal of waste, with a focus on reducing the negative impact of waste on the environment and human health.

Water Management: The process of managing water resources in a sustainable manner, including ensuring access to clean drinking water, reducing water waste, and protecting water ecosystems.

Notes

[i] https://www.unilever.com/planet-and-society/sustainability-reporting-centre/our-sustainability-governance/

[ii] https://www.interface.com/US/en-US/sustainability/sustainability-overview.html

[iii] tesla.com/ns_videos/2022-tesla-impact-report.pdf

[iv] https://www.danone.com/investor-relations/sustainability/reports-and-data.html

[v] https://www.patagonia.com/core-values/

[vi] https://www.ikea.com/ae/en/this-is-ikea/sustainable-everyday/

[vii] https://www.zunocarbon.com/blog/esg-team-structure

[viii] https://www2.deloitte.com/uk/en/insights/environmental-social-governance/importance-of-sustainability-to-employees.html

[ix] https://www.bain.com/about/media-center/press-releases/2023/consumers-say-their-environmental-concerns-are-increasing-due-to-extreme-weather-study-shows-theyre-willing-to-change-behavior-pay-12-more-for-sustainable-products/

[x] https://sponsored.bloomberg.com/article/mubadala/the-future-of-esg-Investing

[xi] https://www.bankrate.com/investing/esg-investing-statistics/

[xii] https://www.esgtoday.com/investors-confront-barclays-at-agm-over-oil-gas-financing/

[xiii] https://www.eco-business.com/press-releases/global-esg-regulation-increases-by-155-per-cent-over-the-past-decade/

[xiv] https://www.weforum.org/agenda/2021/10/no-1-esg-challenge-data-environmental-social-governance-reporting/

[xv] https://www.toyota.com/usa/environmentalsustainability/gri-index

[xvi] https://www.toyota.com/usa/environmentalsustainability/data-report-hub

[xvii] https://www.toyota-europe.com/sustainability

[xviii] global.toyota/pages/global_toyota/sustainability/esg/environmental/climate_public_policies_en.pdf

[xix] https://www.weforum.org/agenda/2021/04/renewable-energy-storage-pumped-batteries-thermal-mechanical/

[xx] https://www.iea.org/reports/deploying-renewables-principles-for-effective-policies

[xxi] .https://ghgprotocol.org/sites/default/files/Guidance_Handbook_2019_FINAL.pdf

[xxii] https://www.apple.com/newsroom/2022/10/apple-calls-on-global-supply-chain-to-decarbonize-by-2030/

[xxiii] https://permutable.ai/apple-esg/

[xxiv] https://about.nike.com/en/newsroom/reports/fy21-nike-inc-impact-report-2

[xxv] https://www.libertymutualgroup.com/documents/lm-esg-materiality-assessment-v3.pdf

[xxvi] https://about.bankofamerica.com/en/making-an-impact/materiality

[xxvii] https://www.coca-colacompany.com/media-center/2030-water-security-strategy

[xxviii] https://services.google.com/fh/files/misc/smb_hub_wellbeing_guidance.pdf

[xxix] https://www.bbc.com/news/technology-53485560

[xxx] https://grow.google/expanding-opportunity/

[xxxi] https://www.businesstoday.in/latest/corporate/story/unilever-pledges-to-halve-use-of-virgin-plastic-by-2025-231888-2019-10-07

[xxxii] https://www.edie.net/ikea-reveals-mixed-progress-towards-climate-positive-and-circular-economy-goals/

[xxxiii] https://www.unilever.com/our-company/our-history-and-archives/2010-2020/

[xxxiv] https://www.unilever.com/news/press-and-media/press-releases/2020/unilever-celebrates-10-years-of-the-sustainable-living-plan/

[xxxv] https://assets.unilever.com/files/92ui5egz/production/452125018d106e2a8025c177f71d986c54b568d6.pdf/uslp-performance-summary-2019.pdf

[xxxvi] https://www.microsoft.com/en-us/corporate-responsibility/sustainability

[xxxvii] https://www.icaew.com/insights/viewpoints-on-the-news/2023/jan-2023/uk-executive-pay-increasingly-linked-to-esg-targets

[xxxviii] https://www.unilever.com/files/deb60070-b400-466b-b2f2-2a567e4c28fc/unilever-2022-ara-directors--remuneration-report.pdf

[xxxix] https://www.nike.com/a/sustainability-2025-targetsA

[xl] https://www.ikea.com/us/en/this-is-ikea/about-us/the-ikea-sustainability-strategy-making-a-real-difference-pubb5534570

[xli] https://qz.com/workers-want-companies-that-care-about-esg-how-to-lever-1849880580

[xlii] https://www.marshmclennan.com/insights/publications/2020/may/esg-as-a-workforce-strategy.html

[xliii] https://www.mckinsey.com/industries/consumer-packaged-goods/our-insights/consumers-care-about-sustainability-and-back-it-up-with-their-wallets

[xliv] https://www.esgtoday.com/consumers-willing-to-pay-12-premium-for-sustainable-products-bain-survey/

[xlv] https://economictimes.indiatimes.com/markets/stocks/news/esg-funds-assets-may-grow-30-every-year-for-a-decade-avendus/articleshow/103916575.cms

[xlvi] https://sustainability-news.net/policy-and-regulation/explainer-csrd-and-what-it-means-for-business/

[xlvii] https://www.unilever-ewa.com/news/2022/were-named-corporate-sustainability-leader-for-11th-consecutive-year/

[xlviii] https://www.greenqueen.com.hk/patagonia-unilever-ikea-among-the-brands-ranked-by-experts-as-global-sustainability-leaders/

[xlix] https://sustainablebrands.com/read/stakeholder-trends-and-insights/unilever-named-in-20th-annual-sustainability-leaders-report

[l] https://globescan.com/2022/06/23/2022-sustainability-leaders-report/

[li] https://www.ikea.com/gb/en/this-is-ikea/climate-environment/the-ikea-sustainability-strategy-pubfea4c210

[lii] Sinziana Dorobantu, Witold J. Henisz, and Lite J. Nartey, "Spinning gold: The financial returns to stakeholder engagement,"
Strategic Management Journal, December 2014, Volume 35, Number 12, pp. 1727–48, onlinelibrary.wiley.com.

[liii].https://www.mckinsey.com/~/media/McKinsey/Business%20Functions/Strategy%20and%20Corporate%20Finance/Our%20Insights/Five%20ways%20that%20ESG%20creates%20value/Five-ways-that-ESG-creates-value.ashx

[liv] Alex Edmans, "The link between job satisfaction and firm value, with implications for corporate social responsibility," Academy of Management Perspectives, November 2012, Volume 26, Number 4, pp. 1–9, journals.aom.org.

lv https://www.iea.org/reports/global-ev-outlook-2023/corporate-strategy

lvi https://www.knowesg.com/featured-article/13-sustainable-designers-and-luxury-fashion-brands-you-need-to-know

lvii https://www.unilever.com/news/news-search/2017/unilever-commits-to-100-recyclable-plastic-packaging-by-2025/#:~:text=Unilever%20commits%20to%20100%25%20recyclable%20plastic%20packaging%20by,industry%20to%20accelerate%20progress%20towards%20the%20circular%20economy.

lviii https://www.apple.com/newsroom/2022/04/apples-self-service-repair-now-available/

lix https://www.allbirds.com/pages/moonshot-zero-carbon-shoes

lx https://sdgs.un.org/goals

lxi https://www.loreal.com/en/commitments-and-responsibilities/for-the-planet/

lxii https://www.kone.com/en/sustainability/

lxiii https://www.unilever.com/news/news-search/2024/unilever-calls-on-industry-associations-to-step-up-climate-efforts/

lxiv https://www.unilever.com/news/press-and-media/press-releases/2020/unilever-celebrates-10-years-of-the-sustainable-living-plan/

lxv https://www.thomasnet.com/articles/other/nike-csr/

lxvi https://www.kingarthurbaking.com/blog/2021/04/22/why-were-committed-to-environmental-stewardship

lxvii https://www.kingarthurbaking.com/blog/2022/04/11/introducing-king-arthurs-guide-to-sustainable-baking

lxviii https://www.ikea.com/global/en/our-business/people-planet/designing-for-a-circular-future/

lxix https://www.ikea.com/global/en/our-business/people-planet/our-circular-agenda/

lxx https://new.oikos-international.org/publications/ikea-and-the-better-cotton-initiative/#:~:text=Through%20its%20partnership%20with%20a%20leading%20environmental%20NGO,mainstreaming%20Better%20Cotton%20as%20a%20new%20market%20commodity.

lxxi https://www.ikea.com/us/en/this-is-ikea/sustainable-everyday/100-committed-to-sustainable-cotton-pub7f285ad1

lxxii https://bettercotton.org/wwf-ikea-release-better-cotton-project-report/

lxxiii https://www.sedex.com/blog/financial-benefits-of-esg-businesses-globally/

lxxiv https://www.mckinsey.com/capabilities/sustainability/our-insights/accelerating-toward-net-zero-the-green-business-building-opportunity

lxxv https://greennetwork.asia/news/how-unilever-transforms-its-business-with-unilever-sustainable-living-plan/

lxxvi https://footwearnews.com/business/business-news/nike-sustainability-2021-impact-report-1203261100/

lxxvii https://edu.google.com/intl/ALL_in/why-google/education-sustainability/

lxxviii https://sustainability.google/

lxxix https://www.investopedia.com/articles/investing/072115/companies-went-bankrupt-innovation-lag.asp

lxxx https://hbr.org/2023/07/the-evolving-role-of-chief-sustainability-officers

lxxxi https://kpmg.com/us/en/articles/2022/tmt-sustainability.html

lxxxii https://www.cbinsights.com/research/report/big-tech-sustainability-climate-tech/

lxxxiii https://corporate.walmart.com/purpose/esgreport

lxxxiv https://corporate.walmart.com/purpose/esgreport/reporting-data/gri

lxxxv https://www.forumforthefuture.org/blog/six-ways-unilever-has-achieved-success-through-sustainability-and-how-your-business-can-too

lxxxvi https://www.unilever.com/planet-and-society/

lxxxvii https://www.unilever.com/planet-and-society/take-action/initiative/volunteer-to-teach-the-next-generation-about-sustainability/

lxxxviii https://www.bankfab.com/en-ae/about-fab/group/in-the-media/fab-launches-frontiers-in-sustainability-executive-education-programme

lxxxix https://www.bbva.com/en/bbva-to-provide-employee-wide-sustainability-training/

xc https://www.patagonia.com/our-footprint/

xci https://eu.patagonia.com/gb/en/sustainable-apparel-coalition.html

xcii https://www.tesla.com/impact/environment

xciii https://www.nestle.pk/csv

xciv https://www.nestle.com/sustainability

xcv https://www.apple.com/environment/mother-nature/

xcvi https://www.linkedin.com/pulse/apples-mother-nature-ad-film-in-depth-analysis/

xcvii https://natureontheboard.com/apples-mother-nature-ad-based-on-a-true-story-cbaedec61fd2 [xcviii] https://www.unilever.com/news/news-search/2023/leading-and-delivering-on-sustainability-through-our-compass-commitments/

[xcix] https://www.unilever.com/planet-and-society/sustainability-reporting-centre/

[c] https://www.europarl.europa.eu/news/en/press-room/20240112IPR16772/meps-adopt-new-law-banning-greenwashing-and-misleading-product-information

[ci] https://www.eco-business.com/news/22-brands-called-out-for-greenwashing-in-2023/

[cii] https://cascale.org/tools-programs/higg-index-tools/

[ciii] https://www.wri.org/publications/corporate-renewable-energy-buyers-principles

[civ] https://clcouncil.org/

[cv] https://www.globalcitizen.org/en/content/us-companies-propose-carbon-tax-to-reduce-co2-emis/

[cvi] https://www.unepfi.org/net-zero-alliance/

[cvii] https://www.tesla.com/blog/tesla-and-panasonic-collaborate-develop-nextgeneration-battery-cell-technology

[cviii] https://sustainablefoodlab.org/

[cix] https://zwia.org/

[cx] https://zwia.org/zero-waste-community-principles/

[cxi] https://fashionforgood.com/museum/about/

[cxii] https://goodfashionfund.com/